Praise for *The Pattern Seekers*

"Based on massive research, Simon Baron-Cohen argues that most of us are specialized in how we perceive the world around us. There are those who focus on people and those who focus on things. The author makes a compelling case that the second kind of mind—the pattern seeker—is at the root of modern human civilization."

—FRANS DE WAAL, author of *Mama's Last Hug*

"Always years ahead of others, always bolder in mind and in action than others, Simon Baron-Cohen now synthesizes a secret of human creativity born out of difference. Where others saw disability, he saw specialness. Like nature itself, the beauty of the human mind comes from its diversity."

—AMI KLIN, Bernie Marcus Distinguished Chair in Autism at Emory University

"A fascinating account of the mechanisms underlying the related capacities of both autistic individuals and innovators."

—BRIAN JOSEPHSON, winner of the Nobel Prize in physics

"*The Pattern Seekers* is a book of big ideas and is sure to excite intense discussion and debate, fueled by Baron-Cohen's lively prose and thought-provoking stories."

—DANIEL J. POVINELLI, author of *World Without Weight*

"Simon Baron-Cohen is one of the greatest thinkers and writers today on the subject of autism. In this erudite new book he explains that autistic people's strongly systematic way of thinking differently is one of the essential elements in the capacity for invention. Baron-Cohen explores how obsessively experimenting with patterns and sequences, whether in music, the visual arts, math, engineering, cooking, or observing the patterns of the ocean waves, led to new inventions and discoveries. He has recalibrated the lens through which autism is understood and redefined it as a rare potentiality, to be valued and celebrated. His bold new idea, that the genes for autism drove the evolution of human invention, places this disability center stage in the story of humans. If you have ever wondered why geniuses spend so much time alone in their sheds, this illuminating book starts to give us an answer to that question."

—JOOLS HOLLAND, musician

"*The Pattern Seekers* is a game-changing book, a passport into exploring the world of innovation and creativity. Most importantly, it celebrates autistic people and is a call for action, to welcome neurodiversity."

—DAVID JOSEPH, chairman and CEO of Universal Music UK

"Simon Baron-Cohen, an internationally acclaimed authority on human brains, has written a fascinating book that illuminates the 'spectrum' of thinking styles. After reading it, you'll better understand the personalities of your friends and colleagues!"

—MARTIN REES, author of *On the Future*

THE PATTERN SEEKERS

THE
PATTERN
SEEKERS

How Autism Drives Human Invention

Simon Baron-Cohen

BASIC BOOKS

New York

Basic Books
Hachette Book Group
1290 Avenue of the Americas, New York, NY 10104
www.basicbooks.com

Printed in the United States of America

First Edition: November 2020

Published by Basic Books, an imprint of Perseus Books, LLC, a subsidiary of Hachette Book Group, Inc. The Basic Books name and logo is a trademark of the Hachette Book Group.

The Hachette Speakers Bureau provides a wide range of authors for speaking events. To find out more, go to www.hachettespeakersbureau.com or call (866) 376-6591.

The publisher is not responsible for websites (or their content) that are not owned by the publisher.

Adapted and redrawn figures are thanks to Patti Isaacs.

Emma Brown assisted with researching and obtaining permissions.

Print book interior design by Jeff Williams.

Library of Congress Cataloging-in-Publication Data
Names: Baron-Cohen, Simon, author.
Title: The pattern seekers : how autism drives human invention / Simon
    Baron-Cohen.
Description: First edition. | New York : Basic Books, 2020. | Includes
    bibliographical references and index.
Identifiers: LCCN 2020019709 | ISBN 9781541647145 (hardcover) | ISBN
    9781541647138 (ebook)
Subjects: LCSH: Autism. | Autistic people.
Classification: LCC RC553.A88 B3684 2020 | DDC 616.85/882—dc23
LC record available at https://lccn.loc.gov/2020019709

ISBNs: 978-1-5416-4714-5 (hardcover); 978-1-5416-4713-8 (ebook)

LSC-C

10   9   8   7   6   5   4   3   2   1   2020

In dedication to autistic people

Sometimes it is the people no one
can imagine anything of who do
the things no one can imagine.

—ALAN TURING
The Imitation Game

Contents

Born Pattern Seekers

Al didn't talk until he was four years old. Even when he started talking, it was clear he was using language differently to most kids. His mind was different right from the start—he was less interested in people and more focused on spotting patterns, and he wanted explanations for everything he saw. He asked people incessant "why?" questions, to understand how things worked. It was exhausting for his listeners. His unstoppable curiosity was at one level refreshing, yet his need for complete explanations was also often just too much for others. He was clearly a different kind of child.

He showed some other unusual characteristics too. For example, he would chant Thomas Gray's "Elegy Written in a Country Churchyard" over and over (a habit that lasted his whole life). At school, his teachers became exasperated with his persistent questioning. One teacher, in frustration, described Al's brain as "addled," meaning confused. But Al's mind was anything but confused. Rather, his relentless questions were requests for greater clarity because he found other people's explanations of how things work vague. He wanted to build up an orderly, evidence-based picture of the world. From his perspective, everyone else's way of thinking was sloppy and imprecise.

But his mother was worried. She could see that her son was frequently being reprimanded in class and put down by his teachers, and she worried this would damage his self-confidence. She needed to act decisively. So, when he was eleven, she decided to pull him out of school completely and to home-school him. This was not a decision she took lightly. But given his insatiable appetite for knowledge and the school viewing him so negatively, this seemed the right thing to do. Her child had a right to learn in the way that suited his different kind of mind.

||||||

Free of the constraints of a conventional school, Al's mother watched with amazement as her son devoured books at home and at the local library. When Al read an account of how something worked, whether it was in chemistry or physics, he would rush down to the basement of the house to conduct his "experiments," to prove that the explanation was true. Free of school, he could finally pursue his passion for seeking patterns in the world, without a teacher telling him to sit still, stop asking questions, and do what he was told. Home-schooling was a liberating gift from mother to son. No longer imprisoned by group learning, Al could finally choose what, when, and how to learn, through individual learning. This suited his mind perfectly, because he was never content to be told by a teacher how something worked, but instead always wanted evidence to verify it for himself. He needed to question all evidence and test things out for himself. His was a mind that didn't follow the crowd. Instead, he wanted to understand things from first principles, to check that his knowledge was *true*.

Al's mother could clearly see that her son's learning style was different. Some described it as pedantic, obsessive, rigid, precise, and exhaustive. For example, when it came to reading in the library, Al would start by reading the last book on the bottom shelf, then systematically read every book in the order they were on the shelves above, not randomly jumping around the bookshelves. He would follow an unbending rule: one book at a time, in a strict, linear sequence, so he could be sure he hadn't missed any information. Even though he was most interested in scientific and technical books, he would never deviate from his rule. And he loved rules, because rules were themselves patterns.

By the age of twelve, Al had read Newton's *Principia*, taught himself physics, questioned theories of electricity, and conducted his own experiments at home to see if they were right. By age fifteen, Al had become fascinated by Morse code, the ultimate language of patterns. And once he became interested in anything, he *had* to master it. He couldn't understand how most people would just dip into lots of topics superficially, since for him a topic had to be understood completely. It was all-or-nothing. He loved how in Morse code the same underlying message could be mapped onto patterns in a variety of ways, using auditory clicks, light flashes, or written symbols. He loved how each letter was a unique sequence of dots or dashes, how a dot was a unit of time, and how a dash was equal to three dots in duration. He loved how a letter was like a musical note, some worth one beat, others two or four beats. He grasped the patterns intuitively—he was a born *pattern seeker*.

When Al was sixteen, he left home. He wandered the country and discovered that his Morse code skills could earn him money, working as a telegraph operator. But at night he would follow his deeper interests, staying up till the small hours, still performing his "moonlight experiments" on whatever machinery he could lay his hands on. Just as when he was a child, he still loved taking things apart to see how they had been assembled, to see what controlled what. And then when he had done that, he was just as excited to reassemble them.

At age just sixteen, Al produced his first public invention. His "automatic repeater" was a device that could transmit Morse code signals between unmanned telegraph stations, so that anyone could translate the code when it suited them. And as we shall see, he would go on inventing right through his adult life.

▌▌▌▌

Two-year-old Jonah was another child who, like Al, was not yet talking. But unlike Al's mother, who stayed calm, Jonah's mother panicked. She was distraught that everyone else's kid was chattering away, so she took her little boy to a pediatric clinic to be assessed.[1]

She watched anxiously as the pediatrician did various tests. The doctor could see that Jonah's mother was worried and thought it might help to show her a chart outlining how every child's language development is different:[2]

> "Can you see how toddlers vary in their rate of language development? They're just different. And which track you end up on depends to some extent on your genes."[3]

Still very upset, Jonah's mother tried to focus on the chart but just couldn't understand it. She explained to the doctor that all the different lines just seemed confusing. She tried to hold back her tears. The doctor put her hand on the arm of Jonah's mother to comfort her as she continued her explanation:

> "You see the solid black line? These are the average kids. And the top line are the early talkers, who are super-sociable, the chatty ones. The bottom line are the late-talkers, who are more spatial, more musical, more mathematical—they love patterns."

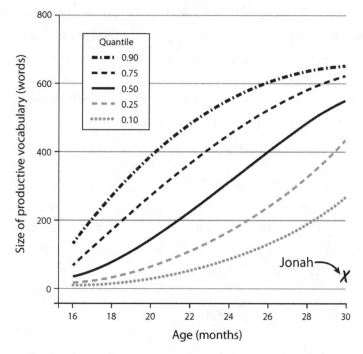

Figure 1.1. Different types of children's language development

The doctor turned to her, waited as if to gauge how much to say, and said:

> "Jonah is one of these kids. They're just not that interested in chatting but are fascinated by how things work. These kids are not better or worse than those on the other tracks. They're just different."

The doctor again paused, and seeing that Jonah's mother was now calming down, she said:

> "I love these kids because they show originality. They may be late to talk, but when they start talking what they say is so much more interesting! Some of them end up as talented musicians

or chess players, some are gifted in math, gardening, cooking, building bicycles, carpentry, or photography. They are perfectionists, who love detail. They spot things that other kids miss."

Jonah's mother was now leaning forward, paying attention to the chart, her tears gone. Then the doctor got out her pen and drew a big X.

"A lot of the kids I see in this clinic are just like Jonah, where the X is, and I've seen them grow up. Some end up as engineers or artists who show originality, successful businessmen or women with a new angle, or scientists who can see patterns in data and make discoveries."[4]

She turned to Jonah's mother:

"And you know what? I was one of these kids. Apparently, I didn't talk till I was three, and I grew into one of those kids who just loved science."

The doctor smiled for a moment, and then looked Jonah's mother straight in the eye:

"Be proud of Jonah. He's just on a different track. Believe me, he'll start talking when he's ready. And if other parents ask why Jonah's not talking yet, just say, 'He's different, but not worse.'"

Just before his third birthday, Jonah did finally start talking. But *how* he used language was unusual. When he spoke, he didn't look up at people to address them. Nor did he use his index finger to *point* at things, to share interest. Instead, he pointed at things to name them, for himself, even when he was alone.

His mother realized that, unlike other kids, he wasn't pointing at things to communicate something about them to another person. Rather, he was pointing to classify objects, for himself. And as he pointed to each object, he named it—endlessly classifying. But she was reassured that at last he was finally talking!

But she noticed another difference in how Jonah used words: He wasn't naming things with general words, like "car" or "mushroom." Rather, he named them with highly specific words, referring, for example, to the precise make, model, and year of a car ("This is a black 2006 Renault Laguna 2.0S"), or the particular species of mushroom ("This is a porcini mushroom").

Jonah's mother was nevertheless very proud of how he spoke, because Jonah's use of language revealed his very *exact* mind, his laser-sharp attention to detail, not unlike her own— she too would notice if the tiniest thing had been moved in the house and felt compelled to always put it back in its original position.[5] It dawned on her that Jonah's language reflected his strong drive to categorize, which was not unlike her husband's fascinations: he would sit for hours poring over books of photographs of different species of birds, or different types of cars. She knew that genes inherited from one or the other parent could cause a child to have blue or brown eyes, but could genes also cause a child to have a mind compelled to be precise and to classify?

She kept in mind the doctor's words: Jonah wasn't worse than other kids, he was just different. She could see that other three-year-olds didn't do what he did. For example, Jonah would sit, rapt, in front of the television watching the weather forecast, to see what had changed in the graphs and the numbers since the last weather report. And still at just three years old, when he was in the hospital for a few days, she noticed he was reading the names of the different drugs on the trolley as the nurse wheeled it past his bed. When she mentioned it to the pediatrician, the doctor called this "hyperlexia," the opposite of dyslexia. Jonah had taught himself to read, even before he had started school. How had this happened? All of her friends needed to sit for hours with their child, painstakingly trying to teach them to read, yet Jonah just took to reading like a duck to water.

One of her friends noticed that whenever she came over to the house, Jonah was always obsessively "experimenting." For example, he would spend hours turning just one light switch in the house—the one at the top of the stairs—to the *down* position, leaving all the other light switches in the *up* position, as though to confirm that the light switch at the top of the stairs controlled the light in the downstairs hallway. He would do this over and over and over again, as if to rerun the experiment, delighted as the light came on, flapping his hands and making a series of high-pitched squeals. When her friend narrowed her eyebrows presumably to say, *What's wrong with him?*, Jonah's mother leapt to her son's defense and said assertively, "Jonah's just different."

By age four, Jonah's interest had moved on to his large collection of toy cars. He would spin one wheel on one car, round and round and round, seemingly getting great pleasure in confirming it spun in exactly the same sequence every time. He would arrange his toy cars in patterns, lining them up in a strict order according to their color and size, and would have a tantrum if anyone rearranged them even slightly.

Jonah's other favorite activity was sitting in front of the washing machine, listening for when it made the precise click or *whirr* he expected to hear as it went through each stage of the cycle. And when it reached a particular predictable point in the sequence, he would flap his hands excitedly. His mother ignored these odd behaviors, feeling they were harmless, and they seemed to make him happy.

But at school the teachers were becoming concerned because Jonah just wouldn't join in. During group reading, when all the kids sat together on the carpet, Jonah would sit with his eyes tight shut, his fingers in both ears. He hated sitting with other children, and he wouldn't look at their faces. The other kids started calling him "finger ears" and would chant it when he came into the classroom, upsetting him. He would run outside when he heard it, leaving his teacher Julia trying to persuade him to come back in. Julia worried about Jonah, and spoke to him gently, asking him how he was feeling. He said he felt anxious when other children moved because they were "unpredictable." She was surprised at a five-year-old using such a grown-up word.

In the playground, Julia noticed that Jonah always tried to keep to himself. Despite the school's best efforts, and even with her help, sometimes he was bullied. She was mortified to discover that on one occasion some kids had picked him up, put him into a dustbin, dumped rubbish on top of him, laughing as he screamed, and then closed the lid. He stayed in there, terrified to move or make a sound in case the bullies were still there, waiting for him to come out. He was in the bin for hours until, fortunately, he was discovered by the school caretaker at the end of the day.

Generally, Jonah preferred to be alone at the edge of the playground, collecting leaves, classifying them into precise categories. Julia, who had by this point decided she should take him under her wing, asked him one day what he was doing. Initially, he didn't answer. When she asked him again, he said, without looking up, in a monotonous voice:

"Yesterday I sorted all the leaves into five different piles: these ones all have a stalk; these ones all have a single blade; these ones all have a smooth edge; these ones all have an elliptical shape; and these ones have a main vein with all the other veins coming off it. But today I realize there's a sixth way leaves can differ: these ones all have leaves that are opposite each other along the stem."[6]

Julia was amazed. She'd never come across a child who was so logical, so different, so self-contained. She asked him why he wanted to find all the different ways to sort the leaves, and he answered simply:

"So I know all the patterns."

Julia felt she was in the presence of a child-scientist who needed no encouragement to conduct his observations but was motivated by pure curiosity to understand the world. When Jonah's mother came to the school gates to collect him that day, Julia told her she should be proud of her son's remarkable mind.

But Jonah's mother was increasingly anxious about his behavior. Other parents were starting to say Jonah was "obsessive" or "weird." He was the only child in the class who wasn't invited to other children's birthday parties. She would dread picking him up at the end of each school day, in case a teacher or a parent came over to her to report yet another incident. On one occasion, Jonah had reacted to another child chanting the "finger-ears" nickname by pushing him so forcefully that the child fell backwards and hit his head. On another occasion, she arrived to pick up Jonah and was called into the head teacher's office. He had apparently picked up some scissors, walked over to a girl sitting at the same table, and cut her bangs because it bothered him that they weren't straight. The little girl was speechless with shock, and her parents were furious.

Jonah's mother longed to have a child who played easily with other children and didn't come home with odd collections in his pockets, of snails, small rocks, or crumpled pieces of paper with his handwritten lists of cars—their make, model, number plate, color, year, and owner—all systematically organized in a grid. And she worried about Jonah because he totally trusted other people.

One time, a child in the playground had asked if he could see Jonah's wallet, and when Jonah agreed and handed it over, the other boy ran off with it. His mother despaired at how she would ever teach him all the different ways someone might trick him. He just didn't seem to understand other children. He said that social interactions were incomprehensible to him, unlike the world of objects or patterns, for which he had an intuitive understanding. So Jonah preferred to be solitary, learning not from others but by and for himself.

It seemed as if everyone completely missed *why* Jonah was doing what he was doing, endlessly sorting and classifying. One child psychiatrist to whom Jonah's mother took him called his behavior RRBI, which he explained stood for "repetitive and restrictive behavior and interests," as if reifying it in this way somehow explained it. To Jonah's mother, calling it RRBI was insulting: the label medicalized his behavior as a symptom of some disease. And she thought it was meaningless because it was totally circular: "Jonah collects things because he's got RRBI."[7]

She decided not to go back to the child psychiatrist and instead to talk to the kind pediatrician, who she felt understood Jonah better. The pediatrician was delighted to see her again and told her that, if you watched Jonah's repetitive behavior carefully, you could see that he was trying to discover the laws for how things work. Jonah's mother felt this doctor was helping her to open her eyes, to see what motivated her son.

And then the pediatrician shocked her:

"I get so annoyed when I hear a psychiatrist calling a child's repetitive behavior RRBI. He may as well say that all science, including medicine, is RRBI. Doesn't he realize that every scientific discovery and every invention that has ever been made over the centuries was discovered through repetition?"

The doctor shook her head.

"When Jonah's doing his experiments, with the light switches, he's like a little scientist, changing just one feature, while trying to hold all the other variables constant, to make discoveries. He's trying to understand the system."[8]

Jonah's mother sat in admiration of this doctor who was helping her to finally see her son as gifted.

||||||||

As young children, Al and Jonah were remarkably similar. They both struggled to understand people, yet their minds were tuned to a hyper level to analyze and understand patterns and systems, questioning, experimenting with, and classifying everything they encountered. Both of these two children, despite being born in different centuries (Al was born in 1847, Jonah in 1988), questioned everything: "Why did X happen? What happens when I do this? Is this an X or a Y? What's the proof that A really causes B and that it's not some other factor C?" With their critical minds, they were constantly analyzing and experimenting.

Both Al and Jonah looked at the world in a fresh way, uninfluenced by social convention, not feeling compelled to follow the consensus. And they both wanted explanations that were complete, without gaps. As his pediatrician had astutely observed, Jonah was like a little scientist, examining every assumption and testing the evidence for it—except that Jonah, like Al, was doing this without any formal training. All these two children seemed to care about was the search for "truth," which for them was simply a word for consistent patterns. Anything that did not fall into a pattern or follow a predictable rule or law was of no interest to them. They were born pattern seekers.

Despite their similar characteristics as children, their lives took very different trajectories. As an adult, Al became famous. He was Thomas Alva Edison, became a celebrated scientist and inventor with 1,093 US patents, and invented remarkable, transformative technologies, such as the lightbulb. He was affectionately nicknamed "the Wizard of Menlo Park" by those who respected his different way of thinking.[9]

In contrast, Jonah today is a young man who still seeks patterns in the world around him. He didn't become a world-famous inventor, but in his own quiet way, he shows the same drive to understand, experiment, and invent. For example, as an adult, he is fascinated by patterns on the surface of the ocean. He drives to the coast every weekend to go fishing, and the local fishermen all know him. Since his teens, they have grown to love having him join them on their boat because, as he gazes at the surface of the sea, he reads the patterns of ripples on the water. The patterns tell him where there's a shoal of fish, how big it is, how deep beneath the surface it is, and even what kind of fish may be in the shoal. Often he says nothing and simply points. The fishermen have learned to trust him, and they throw their nets where he points. They still marvel at how easily Jonah spots patterns they miss. And they say his predictions are always right. The joy that Jonah experiences on these fishing expeditions is palpable, because he can become engrossed in the detail—there's no pressure to see the bigger picture—and these trips also allow him to socialize without having to make conversation.

But even though Jonah has a talent for spotting patterns, remarkable attention to detail, and an extraordinary memory, he has struggled to make a single friend. When I pointed out that the fishermen were his friends, he bluntly corrected me.

"They like me because I show them where the fish are, but after the fishing trips they go to the pub, and I go home alone, and still live with my parents."

Jonah is autistic. But perhaps you already guessed that.

As these two children's stories make plain, the very same behaviors and fascinations can be viewed very differently. Seen through one lens, a child's "obsessions" are a symptom of a "disorder" or a "disease" and associated with disability. Seen through another lens, a child's relentless experimenting and detailed observations are the product of a mind whose pattern-seeking engine operates in overdrive and can lead them to invent, and sometimes to become great inventors.

The capacity to invent is hugely important because once humans became capable of invention, we transformed our world, and we are still doing so today. And yet the capacity to invent is poorly understood. There doesn't seem to be a theory of how we invent, or an understanding of where this transformational ability comes from.[10] The conventional wisdom is that invention involves playing with or exploring an object, seeing it in a new light, or having an insight into it, but these are vague descriptors and don't amount to a theory. Yet when we look at the minds of inventors like Edison, or of autistic people like Jonah, we can glimpse a connection between them that needs exploring.

Glimpsing this connection has driven me to ask some fundamental questions: How do we invent? What goes on in the human mind when we invent? Are humans the only species that can invent? At what point in evolution did we or our ancestors start to invent? What is the intriguing link with autism? And does this link hold true across the autism spectrum, including even those who have learning disabilities or very little language?

As a psychologist and an autism researcher, I have studied the human mind for thirty-five years. In this book, I present a new theory of human invention. Here it is in a nutshell.

First, humans alone have a specific kind of engine in the brain. It's one that seeks out *if-and-then* patterns, the minimum definition of a system. I call this engine in the brain the Systemizing Mechanism. Second, the Systemizing Mechanism developed at a landmark moment in human evolution, between 70,000 to 100,000 years ago, when the first humans began to make complex tools in a way that no previous animal had ever been able to, or any non-human animal can do today.[11] Third, the Systemizing Mechanism allowed humans alone to become the scientific and technological masters of our planet, eclipsing all other species.

Fourth, the Systemizing Mechanism is tuned up super-high in the minds of inventors, and in those in the STEM fields (science, technology, engineering, and mathematics), and in those who strive to perfect any kind of system (such as musicians, craftspeople, filmmakers, photographers, sportspeople, businesspeople, or lawyers, among others). All these people have "hyper-systemizing" minds that can't help focusing on precision and detail and who love to figure out how a system works, how to build a system, or how to improve a system. Fifth, the Systemizing Mechanism is also tuned up very high in the autistic mind. Sixth, the new science shows that systemizing is partly genetic, so it was likely to have been shaped by natural selection. And here's the extraordinary connection: that autistic people, those in STEM, and other hyper-systemizers share these genes.

As we look back across evolutionary time, then at the present, and into the future, we uncover an important truth: those humans who had minds with a Systemizing Mechanism in overdrive were—and are—central to the story of invention.

Chapter 2

The Systemizing Mechanism

When the Systemizing Mechanism evolved in the human brain, between 70,000 and 100,000 years ago, instead of looking at an object (or an event or information) as if there was nothing more that could be done with it, our minds started to look at it as a *system*, something governed by *if-and-then* patterns. The Systemizing Mechanism was the result of a cognitive revolution in the human brain that led *Homo sapiens* to diverge from all other animals and to conquer the Earth. And it all came down to the drive to seek out *if-and-then* patterns.[1]

These three little words are each very special and important, and I need to take you through their meanings carefully. I would ask of you just one thing: when you look at these three little words, don't just assume you know what they mean. *If, and,* and *then* look like three very familiar words in English, but their apparent simplicity masks an underlying depth to each of them. But more on that shortly.

The Systemizing Mechanism entails four steps, which together I call "systemizing."[2]

Step 1 is **asking a question**. When we humans look at the world of objects or events, we start by asking a "why" question ("Why did the candle blow out?"), or a "how" question ("How do birds fly?"), or a "what" question ("What could I do with that piece of wood?"), or a "when" question ("When is it dangerous to go out to sea?"), or a "where" question ("Where is the best place to plant a tomato seed?").[3] There is no evidence that other animals can ask themselves questions, even without words, as we can. It's of course difficult to know if other animals can ask themselves questions, but it's not impossible. That's because an animal or a person doesn't need language to ask themselves a question. For example, it's clear that preverbal children are asking themselves questions when we see them experimenting with a toy in systematic ways to figure out how it works. And it's obvious that a person with no language (for example, after a stroke) can ask themselves questions when we also see them showing curiosity. Indeed, curiosity turns out to be an important indicator of systemizing. Other animals don't show this drive to experiment, and they don't show curiosity, a subject we come back to later.[4] In contrast, human toddlers from age two years onwards constantly ask questions, a sign that they have a Systemizing Mechanism in their brain.[5] And children like Al and Jonah ask questions to the extreme.

Step 2 is **answering the question by hypothesizing an *if-and-then* pattern**. We look for what might have made one thing (the input) *change* to become different (the output). We search the immediate vicinity in case the cause of the change is visible, or we speculate about a cause that must be there but may be invisible. So, if we see a gun barrel (input) that is smoking (output) and the only visible nearby factor that moved was the trigger, we might hypothesize that the trigger being pulled back was the cause of the change. *If* the gun barrel is smokeless, *and* the trigger is pulled, *then* the gun barrel produces smoke.

Step 3 is **testing the *if-and-then* pattern in a loop**. We do this by repeat experimenting, or by making repeat observations, to see if it always holds true. When we test the pattern, this step loops round and round, enabling us to check that we obtain the same results each time. (The loop is depicted by the small black arrow under step 3 in figure 2.1.) The best "systemizers" go around this little loop dozens or even hundreds of times, to be super-sure that the *if-and-then* pattern holds true. If the pattern is confirmed and is new, we have an **invention**.

Finally, in step 4, when we find such a pattern, we **modify the pattern and test the modified pattern in a loop**. We modify the initial *if-and-then* pattern by taking it apart and varying either the *if* and/or the *and* to see what happens to the *then*. We then test the modified pattern by looping round and round to check whether the pattern is seen every time. If the pattern is consistently seen, and if it is new, we have **another invention**. We can then decide whether to keep the modified pattern, either because it improves the efficiency of the system and/or because it has resulted in something altogether new and useful.

Note that sometimes the new *if-and-then* pattern is a **discovery** rather than an invention. For example, when epidemiologists Richard Doll and Austin Hill back in 1954 found that *if* you're a smoker (exposing your lungs to smoke) *and* you smoke more than thirty-five cigarettes per day, *then* you are forty times more likely to get lung cancer,[6] this was a discovery rather than an invention.

In all cases when you systemize, you end up with **control** over the system. Think about sailing downwind: *if* my boat is stationary *and* I hold my sail perpendicular to the wind, *then* my boat moves forward in the same direction as the wind.[7]

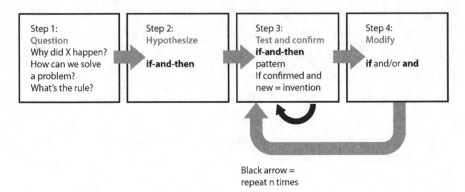

Black arrow = repeat n times

Figure 2.1. How systemizing leads to invention, control, and discovery

The above description may sound a lot like the behavior of a professionally trained scientist or engineer, but let me remind you that we all systemize—we all have a Systemizing Mechanism in our brain. So this theory is not a theory about scientists and engineers, but is a theory about all of us. Though as we'll see, many of those with a highly tuned Systemizing Mechanism may choose to work in a field like science or engineering. Others may learn to master a musical instrument, a craft, or a sport because these too benefit from a strong drive to systemize. The diagram in figure 2.1 looks very abstract, but figure 2.2 shows how the Systemizing Mechanism can be instantiated in a concrete example.

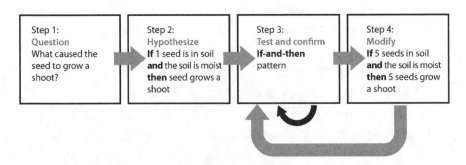

Figure 2.2. How systemizing led to the invention of agriculture

The basics of systemizing are evident in every young child, and in all of us as we ask ourselves questions and try to figure out how things work. It's evident in watching a young child exploring an object, as they discover what can be done with it or what it can do. And that playful *curiosity*—driven by wanting to understand a system or wanting to solve a challenge—*is* everyday systemizing at work. It's that "Aha!" moment, when a toddler figures out how to balance a tower of toy bricks, or how to control a water faucet to change the flow of water, or how to push a light switch to turn on a light. *If-and-then* thinking.

When we identify an *if-and-then* pattern, that is a system. Hence my choice of the word "systemizing." And it turns out that every tool is a system, from the earliest bow-and-arrow to the first musical instrument to the modern-day text message. These are all just tools that we invented to do work for us. My claim is that our only route to inventing a tool, or improving on an existing tool, is via the Systemizing Mechanism and its four steps.[8]

Just to give you a flavor of how the Systemizing Mechanism works, here's one of my favorite examples of a mechanical system, invented about 5,000 years ago to address a big question: How can a massively heavy object be moved? Back then, someone must have looked at a heavy stone, for example, and then looked at their ox for its potential use in an *if-and-then* pattern. What I like about this example is that the ox already existed, but a human looked at the ox in a new way: *if* a stone is hugely heavy, *and* I harness it to my ox, *then* the huge stone moves. The ox was no longer just an ox but now was seen as a causal operation in an *if-and-then* algorithm. Historians think this is how the huge stones at Stonehenge in England were transported to their eventual location 5,019 years ago.[9] This invention likely built on an earlier new tool, the wheel, invented about five hundred years earlier.[10] Combining these two inventions yielded the ox pulling the hugely heavy stone on rollers, or on a sledge.[11] This, like many inventions, was scalable (figure 2.3).

Figure 2.3. Inventing a way of moving a heavy stone

Let's go back to those three little words—*if*, *and*, and *then*—to briefly unpack their precise meanings, since my whole thesis revolves around them. (I joked to my editor that this book could be the shortest book in the world—just three words long. Sensibly, she asked me to elaborate on them.)

The word *if* has at least three meanings: as a hypothetical, as in "if X is true"; or as an antecedent, as in "if X happens first"; or just to denote the input, as in the starting state of an object or event.

The word *and* is like a magic word because it refers to an operation, like addition or subtraction, or something that is done to the input. At its most powerful, *and* refers to a **causal** operation.[12] I think of a causal operation as magic because it changes one thing (the input, the *if*) into something different (the output, the *then*). Think of the *and* in the following pattern: *if* ice is in a bowl, *and* the bowl is on a flame, *then* ice turns to water. The *and* is the cause of the change.

Finally, the word *then* also has at least three meanings: as the conclusion, as in "then Y is true"; or as the consequence, as in "then Y follows"; or just to denote the output, as in whatever the input was transformed into. This also means that *if-and-then* patterns can also be described as "input-operation-output" patterns (figure 2.4).

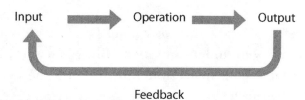

Feedback

Figure 2.4. One meaning of *if-and-then*. These three little words map onto what engineers call "input-operation-output."

Humans alone systemize, and we do this to discover, to solve, to control, and to invent.[13] When the Systemizing Mechanism evolved, it meant humans could not only invent a tool but also see existing tools in a new light: we could understand a tool, then make small changes to it to produce a new, potentially better tool. The *if-and-then* algorithm enabled humans to ratchet up earlier tools to create ever new variations, new tools. And as humans continued to do this, rerunning the algorithm, each time tweaking the *if* and/or the *and* variable, this led to runaway invention.

Today we are surrounded by a myriad of complex tools in our everyday lives, many of which we just take for granted, but that someone at some point invented. I call them "complex" because they are more complex than the ones we see non-human animals using. Yet they're not necessary hugely complex. Some are so mundane—like the fork you eat your salad with, your coffee cup, or even the chair you are sitting on—that you may not even think of them as tools, but they are. These are all *mechanical* systems. Some are less mundane because they led to a big advance, like the glasses sitting right on your nose, which effectively cured vision problems. And some of these mechanical systems were game-changing, like human aviation, the result of experiments by Sir George Caley, the nineteenth—century British inventor who discovered that *if* a glider has a fixed wing, *and* the wing has an angle of incidence of six degrees, *then* the glider will be lifted up.[14]

When we systemize, we look for systems (*if-and-then* patterns) in the world to try to understand them. But mechanical systems, like the ox pulling the heavy stone, are just one variety of system. When we gaze at the world, we also see a myriad of *natural* systems: the changing weather as snowflakes fall, the movement of the wings of a dragonfly, the motion of tidal waves—all amenable to an *if-and-then* pattern analysis, and which we can use to make predictions: *if* there are cumulonimbus clouds in the sky, *and* there is thunder, *then* there will be severe weather.[15] Such a prediction can be a very useful early-warning system.

With other natural systems, once we have analyzed the *if-and-then* rule, we can control the system, as when humans invented agriculture. (For example, *if* a tomato seed is in the soil, *and* the soil is moist, *then* the seed will grow into a tomato.) The agricultural revolution 12,000 years ago transformed how humans lived, from a nomadic hunter-gatherer lifestyle that enabled us to feed our families, to the first farms that then enabled us to feed a whole village.

Or consider another natural system. The invention of medicine, unique to humans, answered specific questions, such as "Why did my headache go away?" Herbal medicine is at least 3,300 years old and must have begun when someone hypothesized, for example, that "*if* I have a headache, *and* I eat willow bark, *then* my headache goes away."[16] Why don't we see apes or monkeys experimenting with herbs when they're ill? We'll look at non-human primates and other animals later in the book, including whether they self-medicate, but in short, I will argue that non-human animals just don't experiment.[17] In human societies, for thousands of years, the person who experimented with the effects of plants on health was often elevated to the status of a healer. Modern society still elevates our best medical researchers, deservedly, to a high status because if the *if-and-then* rule they have discovered holds true, then it has the power to treat a disease across the whole population, at scale, not just in one patient. But all that these medicine men and women are doing, just as they did thousands of years ago and still do today, is just carefully checking *if-and-then* patterns: "*if* I measure the size of a cancer tumor, *and* add this specific drug, *then* the tumor shrinks."

Once we analyze the *if-and-then* rules governing a system, we understand how the system works. Of course, often there's more to learn, but systemizing is iterative—we can keep exploring the system further to learn more about its inner workings, but the method is always the same: *if-and-then*.

Harnessing the new knowledge that comes from the Systemizing Mechanism enabled humans to discover how natural systems such as the kidney work and allowed us to invent mechanical systems such as the windmill, the microscope, and the telescope. Today we can breed a new flower, edit a gene, design a new medical drug, or build a new hospital—all systems of different orders of magnitude, all under our control, and all the product of the Systemizing Mechanism. This humble mechanism in the human brain has been grinding away for 70,000 to 100,000 years, delivering ever more impressive inventions.[18]

||||||

The nineteenth-century English mathematician George Boole first described *if-and-then* thinking in his analysis of logic, and I acknowledge I am borrowing his terms to describe the workings of the Systemizing Mechanism.[19] He is credited with inspiring the invention of modern electronics, the development of the modern computer, and the digital revolution, but to my mind an equally important legacy is the clear terminology he left us for describing the logic of the Systemizing Mechanism. In describing logic in terms of *if-and-then*, Boole's terminology beautifully captures the essence of systemizing. To give him credit for this insight, we could call systemizing the "Boolean mind," which I argue is exclusively human.

The son of a shoemaker, Boole left school after primary school. He was taught by his father and was otherwise largely self-educated. He ended up a mathematician and philosopher of logic and in 1854 wrote a book called *The Laws of Thought*. To support his siblings and his parents, as the only breadwinner in his family, he became a teacher at age sixteen in Doncaster in Yorkshire. Impressively, by age nineteen he had opened his own school in Lincoln in the East Midlands. Fifteen years later, and despite his lack of any formal mathematics education, he was appointed professor of mathematics at Queens College in Cork, Ireland.

One November day in 1864, aged just forty-nine years old, Boole walked the three miles from his cottage to the university, in the pouring rain, to give his lectures. He arrived soaked through. When he got home, he developed a fever, and his wife Mary, using a bizarre form of logic derived from the homeopathy she practiced, thought that the remedy for an illness should resemble its cause. She therefore wrapped him in damp sheets—some accounts say she also poured buckets of water over him. Poor George got worse (unsurprisingly to us, as modern readers), and he died a few days later.

It is a twist of tragic irony that Boole, the remarkable logician, met his end through Mary's faulty logic. (It is doubly ironic that Mary was an accomplished mathematician in her own right.[20]) I've not read any account accusing Mary Boole of inadvertently killing her husband, but it could be construed that way. Clearly her intention was to heal him, but his death at just forty-nine years of age deprived the world of one of the greatest logical minds. Fortunately, he had already made an enormous contribution. His intellectual legacy is reflected in the branch of algebra named after him (as is a crater on the moon).

Even two-year-old human children can systemize using basic *if-and-then* logic.[21] This is a clue that we (and no other species) are partly innately *wired* to seek out these patterns. Preschool children ask why a novel object behaves unexpectedly, and they look for explanations (causes). Even more impressive is that they run "tests" to figure out what's causing what, keeping track of when there are anomalies, and can identify different kinds of causal chains. For example, preschool children can distinguish between different mechanisms—how a switch might drive one cog wheel rather than another, or drive one cog wheel to drive another—and as they do so they look for evidence that supports different *if-and-then* patterns.[22] So systemizing seems partly hardwired into the human brain.

||||||

There are three different ways we test an *if-and-then* pattern. The first is via **observation**. We often resort to observation when the pattern is so big that we can't get our hands on the variables to manipulate them. An example would be trying to understand and answer the question "Why does the moon's shape appear to change?" Observation is a powerful way to spot a change and start to hypothesize about and test for an *if-and-then* pattern. Observation can of course also be used to answer a question about a small-scale, highly delicate phenomenon, such as watching a spider in order to answer the question: How does a spider navigate her web? When we look back at the lives of great scientists, many often spent their playtime as children observing all kinds of patterns in nature out in the back garden, just as Jonah did.

The second way we systemize is via **experimentation**. We usually resort to experimentation when we *can* get our hands dirty and test a system, such as trying to figure out and answer the question: How does the kitchen toaster work? Or looking at a

Venus fly trap, we might try to answer the question: How does the Venus fly trap work? We test our hypothetical *if-and-then* pattern by pretending to be a fly, delicately poking its leaves using a little twig to reveal the key mechanism: when one of the tiny short stiff "hairs" on the leaves is bent, that motion triggers the two lobes of the leaves to snap shut in less than a second.[23]

Finally, the last way we systemize is via **modeling**. Modeling is doing an experiment using a smaller or simpler model, such as building a model of a bridge that fits on a tabletop and then scaling it up, so the bridge might end up being as long as the one that links Denmark to Sweden.

Let's look at each of these ways of systemizing in more detail.

Systemizing via observation of *if-and-then* patterns is what we do whenever we are out in nature. Consider the story of Papillon, the penal colony prisoner who sat on the cliff top and watched the sea tide for weeks. The big question he was faced with was: How can I escape this remote island prison? He had all day every day to think about it. Gazing out to sea, and based on the patterns he was observing, he hypothesized that "*if* I jump strapped to a homemade raft, *and* land on the seventh wave, *then* the wave will be strong enough to carry me safely out to sea." Systemizing the tidal movements of the waves saved his life when he finally jumped and escaped to freedom, since landing on any other wave would have led to certain death by being dragged back onto the rocks.

But early humans weren't just looking at the waves. They also loved to systemize the night sky, just as we like to do today when we are out camping or sitting around the fire with friends in the garden on a summer's night. Imagine an observer, 10,000 years ago, lying on her back at night and asking herself a big question: What's causing the shape of the moon to appear to change? She notices that *if* it's day twenty-nine since the last full moon, *and* I look at it on day thirty, *then* the moon's shape will become a perfect circle. The Systemizing Mechanism works by locking on to a problem and trying to see the pattern in order to understand it.

The Systemizing Mechanism also drives our stargazer from 10,000 years ago to do something very odd, something you never see in other animals. She looks at the moon from exactly the same place every night in order to understand one system: the "moon-as-seen-from-this-particular-spot." In doing this, she is conducting systematic observations. Why don't monkeys do that? Our human observer's behavior is driven by her Systemizing Mechanism, which in turn drives her *curiosity, for its own sake.* The reward is intrinsic: the pleasure of her curiosity being satisfied and confirming the repeating *if-and-then* pattern. Curiosity *is* the grinding of the Systemizing Mechanism.[24]

We know from archaeology that humans as long as 10,000 years ago were indeed systematically logging the changing shape of the moon. Archaeologists in Scotland found what they believed to be a lunar calendar, comprising a series of twelve specially shaped pits designed to mimic the phases of the moon. The pits align on the midwinter solstice in a way that would have helped their makers keep track of the lunar cycle.[25] And sky-gazing was also recorded 3,500 years ago, by the Sumerians in Mesopotamia, who named many of the stars.[26] In the Babylonian era, in 164 BC, humans were recording observations of what we now call Halley's Comet.[27] So humans, using their Systemizing Mechanism, were systemizing the skies long before "science" was formally invented.

Now imagine our sky-gazer is a man in China, over 3,000 years ago, who is curious not just about why the shape of the moon is changing, but also about what is causing the *color* of the moon to appear to change. There was such a man, who wrote a book called the *Zhou Shu*, which was discovered in a tomb in AD 280.[28] The lunar event he described is thought to have taken place in 136 BC, over 2,000 years ago.[29] Seeing the moon appear to change from whitish gray to luminous red is breathtaking.

Today we know that the color of the moon appears to change when a lunar eclipse occurs, and that the *Zhou Shu* describes what today we would call a lunar eclipse. Here's the systemizing algorithm: "*if* I observe the white moon, *and* it lies in a straight line with the sun and the Earth, *then* the moon will seem red." Our Systemizing Mechanism enables us to reveal the secret laws of how the universe works, secrets that apes, monkeys, and other species will never know or understand.

▌▌▌▌▌

If there is one person whom we associate with systemizing the natural world, it's Swedish botanist and zoologist Carl Linnaeus. You likely know of him as the father of modern tax-onomy. Admired for his detailed, hierarchical classification of animals and plants, Linnaeus was a *hyper-systemizer*, meaning that he systemized unstoppably, compared to most of us.[30] We'll come back to the notion that the Systemizing Mechanism can be tuned to different levels, that in most of us it's tuned at an average level, but that in people like Linnaeus it's tuned super-high. (Carl was not the only hyper-systemizer in his family: for example, his brother Samuel wrote a manual on beekeeping. We'll also return to the question of whether hyper-systemizing runs in families.) By the age of seventeen, Linnaeus had become a voracious reader of all the major books on plants. And in 1732, at just twenty-five years old, he embarked on a six-month sample-hunting expedition to Lapland.

Like Darwin during his famous trip to the Galápagos, Linnaeus traveled by foot and on horseback, covering over 1,000 miles and gathering hundreds of samples of plants, going through each specimen to identify shared and distinc-tive features. His observations culminated in the publication of *Flora Lapponica*, a taxonomy of 534 distinct species. Then, whenever any new specimen was found, he would review his existing classification system to see if it fit, and if it didn't, he created a new taxonomic category for it.[31] Linnaeus was the eighteenth-century equivalent of young Jonah in the playground, unstoppably classifying nature.

His grand and ambitious goal was to be able to tell apart any two classes of plants, even those that looked superficially alike, and he succeeded. In 1735, he published a book called *Systema Naturae*, initially just twelve pages long, as a proof of principle of how to classify and systemize nature. After it was published, readers started sending Linnaeus more specimens to be classified and included. By 1758, when *Systema Naturae* reached its tenth edition, it included 4,400 species of animals and 7,700 species of plants. Linnaeus's Systemizing Mechanism was clearly tuned super-high. As we'll see, those with a strong systemizing drive soak up a huge amount of information and organize it systematically into the *if-and-then* patterns they detect in the world.

||||||

To create a classification system of this kind requires checking off each item on a mental or an actual list, finding the rule that uniquely defines the features of a particular plant or animal. Such taxonomies require identifying the *if-and-then* rule (for example, "*if* the specimen has a black head, *and* a red belly, *then* it's a robin"). Think how Jonah was carefully classifying each leaf in the playground. Bird-watchers and their modern-day equivalents, such as plane-spotters or train-spotters, will go out for hours to seek these patterns, come rain or shine.[32] They too are hyper-systemizers.

The very same Boolean logic allows us to categorize *any* living thing, such as the 7,500 different species of apples that exist in the world: "*if* it's got an all-green skin, *and* tastes tart, *and* has a hard feel *and* a crisp bite, *then* it's a Granny Smith." (You can add as many *ands* into the algorithm as you like, to make it more and more precise, so long as it still fits the *if-and-then* algorithm.) How we distinguish the varieties of apples in the supermarket today is how early *Homo sapiens* distinguished other species of food in the forest or on the savannah plains 70,000 to 100,000 years ago.[33]

My contention is that pre-modern humans, like non-human animals today, could see the apples but did not systemize them. That's because, as we'll see when we look at their tool use, there is no convincing evidence that our hominid ancestors could systemize. Of course, just like apes and monkeys today, our hominid ancestors knew an apple was edible. They could also discriminate apple A from apple B, learn that A might taste good and B might make them nauseous, and so could develop food preferences and aversions. Equally, just like apes and monkeys today, our hominid ancestors could also see the moon was changing shape or color.

But critically, as we'll see later, there is no robust evidence that hominids before *Homo sapiens* could spot *causal* patterns. It's not that our ancestors before 70,000 to 100,000 years ago were incapable of recognizing all patterns; they could certainly recognize simple patterns such as "A is associated with B." Even a rat or monkey can do this, using "statistical learning": keeping track of how likely it is that A will be associated with B.[34] And a rat or monkey can also use other pattern recognition processes, such as associative learning, which is particularly powerful when B is a reward or a punishment.[35]

Simple associative learning of this kind can explain how our hominid ancestors produced simple tools—like using a rock as a hammer to crack open a shell to get the nut inside, or using a stone ax to cut and scrape—and how they learned what was good to eat. They may even have used spears as a weapon. But in contrast to modern humans, and as so beautifully magnified in Linnaeus, it's unlikely that our hominid ancestors *systematically* sorted apples or indeed systematically sorted anything, into categories, or observed *if-and-then* patterns, or experimentally tested for them. Equally, there's no convincing evidence that modern apes or monkeys do these things.

When you look at the lightning speed of strong systemizers among humans today and ask how they are able to spot patterns and recall facts so rapidly, it becomes evident that they are building **mental spreadsheets**, using *if-and-then* patterns where *if* is the row, *and* is the column, and *then* is where they intersect.[36] We can systemize objects or events in space (*where* we saw it) and time (*when* we saw it), and as we collect more data or examples, the patterns or laws emerge. By systemizing the natural world, early systemizers would thus have derived the laws of nature—knowledge, for example, about where and when to plant flowers in the garden. Here's just one "where" example: *if* you take a rhododendron, *and* plant it in alkaline soil, *then* its blooms will change color.[37] You can imagine just one mental spreadsheet of a gardening expert: each row might be a particular plant name, each column might be the soil type, and where they intersect might be the color of that plant's flowers.

If you're in any doubt that humans love to systemize plants, just visit the Chelsea Flower Show in London. The fact that the event sells out very quickly shows how the human mind loves to systemize plants, and the visitors are not all professional garden designers. Of course, some people go to the show just to enjoy the beauty of garden design, but many go to understand the plants, using *if-and-then* logic to learn when and where they can be planted, and with what effect.

We can systemize anything, from the menstrual cycle and its link to fertility, to predicting when volcanos will erupt, to classifying rocks by how they were formed.[38] Consider how Sir Isaac Newton inferred gravity as a cause from seeing an apple fall from a tree (in my college, Trinity, in Cambridge): *if* an apple is unsupported, *and* there is a gravitational force, *then* it will fall toward the Earth.[39] The ocean's tides are another aspect of nature we systemize via observation. The discovery that tidal patterns are caused by the moon, for example, was documented at least 3,000 years ago.[40] The tide information available at the seaside is usually in the form of a table or spreadsheet, reflecting the systematic organization of the information in the mind of a strong systemizer. And when surfers are out in the ocean, they aren't just observing the tide. They are also classifying waves by their shape and predicting what they will do. These examples illustrate how a single algorithm—*if-and-then* reasoning—allows us to explain an infinite and diverse range of natural phenomena.

▌▌▌▌

The second way we systemize is by **experimenting**. To me, this was the real payoff of the evolution of the Systemizing Mechanism—the ability to understand how things work by identifying causal patterns through experimentation. Think of Al as a young child, performing his chemistry experiments in the basement of his house, or think of the young Jonah, repeatedly (some would say obsessively) flipping the light switch from up to down to see the causal effect on the light going from off to on. Experimenting is not just what scientists, engineers, mechanics, doctors, musicians, cooks, and craftspeople do. It's what humans have been doing for 70,000 to 100,000 years. It's all about taking step-by-step actions, monitoring the results of our *if-and-then* reasoning.

To dispel the notion that it's only scientists who systemize via experimenting, consider how even in the kitchen, when you boil an egg, you are systemizing by experimenting. *If* I take an egg, *and* I boil it for eight minutes, *then* the yolk will be hard and yellow. *If* I take an egg *and* I boil it for four minutes, *then* the yolk will be soft and orange.

When I was a young child, my brother and I used to love going to the playground, to sit on the seesaw. We would play around to see how high we could go. Unknown to us, we were doing what all kids do on a seesaw: systemizing via experimenting. "*If* my feet are on the ground and my brother is off the ground, *and* I push up with my feet, *then* I go up and my brother goes down." Why don't monkeys and apes build seesaws in the wild? Why don't they spend hours playing with seesaws, to figure out how to control the system? They just aren't interested in *if-and-then* patterns. If we ever see two monkeys build a seesaw and play on it, we should worry about humans' primary position on the planet, because it would mean they have evolved a Systemizing Mechanism in their brain, and they could start to invent.

The seesaws of my childhood could only go up and down, but last week, walking through the park, I noticed that someone had invented a different kind of seesaw. There were two kids, sitting on opposite ends of a heliocyclic seesaw, giggling unstoppably as they played with motion. This new seesaw was not the old-fashioned type that was fixed in the center with a pivot, so that all you could do was go up and down. The new seesaw instead sat on a ball-and-socket, so that the plank could move in any plane in three-dimensional space, going round and round, rotating in a hundred different directions. Someone had changed the *and* variable in the seesaw. The Systemizing Mechanism had given rise to an invention.

Systemizing via experimenting is also part of how we do sports. Consider skateboarders "dropping in" to the half-pipe, shifting their weight to turn corners, grinding along the edge of a step, getting some "air" by traveling at high speed up a near-vertical wall, doing a somersault, and landing on the wheels to continue skateboarding. Some of these are just young kids who spend hours in the front yard on the mini-ramps they've created, shifting their weight from one foot to another to create the desired outcome as they skateboard down them. They are flipping the skateboard 180 degrees, making the skateboard spin, or driving the skateboard upwards to jump up onto a park bench. None of these impressive stunts would be possible without the skateboarder having systemized each move, doing the maneuver over and over and over again, spotting and performing the repeating patterns, until it is perfected, and coming up with new moves that impress the audience. It turns out that there are at least three hundred *if-and-then* moves you can do with a skateboard, and human teenagers (largely boys) engage in what seems like endless repetition in our town parks, or on the smooth concrete of the streets and plazas of many cities, to enjoy a skate.[41] They are experimenting, in this case, with their own movement. Gymnasts and other sportspeople do the same.

The third and final way we systemize is via **experimenting within a model**. A clear example was Alexander Fleming's famous discovery of penicillin's antibiotic properties. He used a Petri dish with bacteria in it as a model of a human wound containing bacteria. He did this because an actual wound is way too messy, damaged, complex, uncontrolled, and layered to see what is causing what. A model simplifies the real world and does so on a manageable scale you can work with.

The big question Fleming asked was: How can we stop soldiers in World War I from dying of sepsis? At that time, the treatment for infected injuries was antiseptic dressings, but antiseptics, he observed, were actually killing more soldiers than infection itself. He speculated that the antiseptics often made the injuries worse because, although they killed bacteria on the surface of the wound, bacteria could still shelter deep in the wound.

Fleming was studying the bacterium *staphylococcus* in his model environment (the Petri dish) at St. Mary's Hospital in London. Before going on holiday with his family in August 1927, he stacked all his Petri dishes on a bench in the corner of his lab. On his return to the lab, on September 3, he noticed that one culture was infected with a fungus. In particular, he observed that the colonies of the bacteria immediately next to the fungus had been destroyed, while those farther away were unaffected. He hypothesized that "mold juice" must be killing the bacteria: *if* there is a living bacterial colony, *and* it is close to mold juice, *then* the bacteria will die. The fungus was penicillin, and he had inadvertently discovered the first-ever antibiotic. As a result of his systemizing, Fleming received the Nobel Prize in Medicine and Physiology in 1945. He wrote, I think with charming modesty,

> One sometimes finds what one is not looking for. When I woke up just after dawn on September 28, 1928, I certainly didn't plan to revolutionize all medicine by discovering the world's first antibiotic, or bacteria killer. But I suppose that was exactly what I did.[42]

This famous episode is often cited as an example of how a discovery can be made by serendipity, but it fits the Systemizing Mechanism beautifully. In step 1, Fleming had a big question, and in step 2 he hypothesized an *if-and-then* pattern via observation of a chance finding. In step 3, he tested and retested this hypothesis to confirm his novel finding. Others went to step 4 to modify the pattern (for example, by scaling up production), but looping step 3 alone explains both the discovery and its confirmation.

▌▌▌▌▌

How is systemizing implemented in the human brain? Neuroscientist Mike Lombardo and I examined brain-scanning (functional magnetic resonance imaging, or fMRI) data from studies in which a person is asked to do tasks related to systemizing. We looked at tasks involving attention to detail,

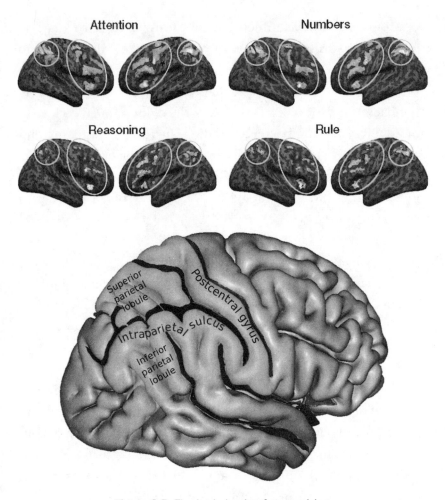

Figure 2.5. The brain basis of systemizing.

Top: Key brain regions supporting mental processes relevant to the Systemizing Mechanism.

Bottom: The intraparietal sulcus, active during numerical and mechanical reasoning.

error checking, reasoning, rule learning, numerical reasoning, and pattern recognition. All these aspects of systemizing use the sensory-perceptual areas of the brain. This makes sense because, when you systemize, you analyze the detail coming in through your senses, telling you what the world is actually like. In addition, the fMRI studies show that the Systemizing

Mechanism depends strongly on lateral frontoparietal connections within the brain and a brain region called the intraparietal sulcus (IPS).[43] The IPS is also active when humans engage in tool-making and is atypical in children with dyscalculia (those who struggle to systemize numbers). While more experiments are needed to determine the brain basis of systemizing, these already give us an indication of how the brain implements the Systemizing Mechanism.[44]

||||||

Can we also systemize aspects of people's behavior? Clearly it is hard to see rules or patterns in one-off behaviors—why someone said what they did, or why someone's emotional expression changed so fast—but we can systemize repeating behavior, such as someone trying to perform the same sequence of actions every time.[45] Examples might include a fisherman casting a fishing line, a golfer swinging a golf club, a dancer performing a dance routine, a gymnast executing a somersault on a trampoline, a basketball player shooting a basketball hoop, a congregation repeating a ceremonial chant, a singer singing a song, or a guitarist playing a riff on a guitar. Because such actions can be repeated again and again, to perfect the action sequence, the Systemizing Mechanism can latch onto these "motoric" patterns just as easily as it does when confronted with a pattern like a railway timetable, a light switch, or the changing shape of the moon. We can also systemize some social systems, like the chain of command in a military unit, the rituals in a religious or military ceremony, the processes in a business, or adherence to moral and legal codes.[46]

We can even systemize a novel or a drama, seeing each word or phrase in the writing as a tool designed to do work for the writer, or each scene functioning to progress the drama in more unexpected or engaging ways. And we can systemize the characters, their actions, thoughts, and emotions, again as tools in the overall success of the narrative. We can do this because in a novel the moving parts are entirely scripted, the narrative sequence will be the same over and over again, the writer or editor can position each word, each utterance, each glance, and each action and reaction. This is like a static jigsaw puzzle with a thousand pieces, and there's a logical way to fit them all together in the best possible way. The novel or the movie or the theater script is social behavior, frozen into a script that can be routinized and rehearsed and experimentally manipulated to optimize its performance, just like any complex tool.

But most of our everyday behavior is not lawful in this sense: when we chat with friends, we don't have the very same conversations every time. And systemizing other people's thoughts and feelings will inevitably fail: when we experience emotions, they don't bubble up in exactly the same way, following the same triggers.[47] Nor do our beliefs stay the same as they are changed by our experience, and nor do our relationships stay the same across time. So systemizing fails us in most of our social interactions. Indeed, most of us don't depend on systemizing to navigate the social world. Instead, we use another game-changing mechanism in the human brain, the Empathy Circuit.

I hope you agree that the evolution of the Systemizing Mechanism has been remarkable in causing what I and others call the *cognitive revolution*, a step change in our understanding of the world and enabling our capacity to invent. But the Systemizing Mechanism was not the only astonishing change in the human brain: the Empathy Circuit, a second and, again, uniquely human brain mechanism, also led *Homo sapiens* to diverge from all other animals. The Empathy Circuit was the subject of my earlier books *Zero Degrees of Empathy*, *The Essential Difference*, and *Mindblindness*, so I do not examine it at length here. But in short, the Empathy Circuit allowed us to think about others' thoughts and feelings and to think about our own thoughts and feelings, rapidly, second by second, in real time in a dynamic social context. By enabling us to imagine other people's mental states (their thoughts, feelings, intentions, and desires), not lawfully but flexibly, we could anticipate what they would be likely to do next, in real time, and to react to their thoughts and feelings rapidly with an appropriate emotion of our own.

The Empathy Circuit in the modern human brain has at least two networks: one supporting *cognitive empathy*, defined as the ability to imagine the thoughts and feelings of another person or animal; and the second supporting *affective empathy*, defined as the drive to respond to another's thoughts and feelings with an appropriate emotion. Cognitive empathy is the recognition element, while affective empathy is the response element. Cognitive empathy is what primatologist David Premack called having a *theory of mind* and is what enables modern humans to navigate the social world. This is because cognitive empathy allows us to imagine what another person may be thinking or feeling about things in the world, and particularly what another person may be thinking about *us.* Having a theory of mind also means that you can reflect on your own thoughts and feelings—it enables self-reflection. Non-human primates such as chimpanzees, and likely other animals, may have elements of a theory of mind, such as being able to recognize another animal's goals and desires, but there is still no strong evidence that, unlike us, they can imagine another animal's *beliefs.*[48]

The Empathy Circuit resides in a network in the human brain of at least ten brain regions, including the ventromedial prefrontal cortex and the amygdala.

When we think about the cognitive revolution, with new brain mechanisms enabling humans to do things that their

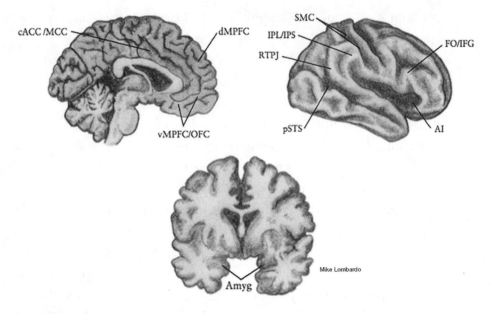

dMPFC = dorsal Medial Prefrontal Cortex;
vMPFC = ventral Medial Prefrontal Cortex;
OFC =   Orbito-Frontal Cortex;
FO = Frontal operculum;
IFG = Inferior Frontal Gyrus;
cAAC = caudal anterior cingulate cortex;
MCC = middle cingulate cortex;

AI = Anterior Insula;
RTPJ = Right Temporal-Parietal Junction;
STS = Superior Temporal Sulcus;
SMC = Somatosensory Cortex;
IPL = Inferior Parietal Lobule;
IPS = Inferior Parietal Sulcus;
Amyg = Amygdala

Figure 2.6. The Empathy Circuit

ancestors could not, we need to keep in mind two things. First, that evolution typically works via incremental tiny changes, rather than quantum leaps. And second, that natural selection of traits tends to work via hundreds if not thousands of common genetic variants, each having a tiny effect but contributing to individual differences in a trait, rather than a single gene leading to the trait being either present or absent. We'll come back to the genetics of empathy and systemizing, but so it is with the Systemizing Mechanism, of which there were likely to have been *proto* or primitive forms, such as statistical learning and simple tool use, as seen in many other animals. Equally, so it is with the Empathy Circuit, of which there were likely to have been proto forms as well, such as responding to another's distress calls, also as seen in many other animals.

But what we see in humans and no other species is the *full* Systemizing Mechanism, which drives the curiosity to ask questions and to experiment with variations in patterns, and the *full* Empathy Circuit, which includes a theory of mind. To see quite how revolutionary the Empathy Circuit was in the evolution of the human brain, it is useful to recap the main advantages conferred on us by a theory of mind. If someone has a theory of mind, and particularly the ability to imagine another's beliefs, they should be capable of at least three skills that would have massively transformed their lives.

First, a theory of mind allows for *flexible deception*, or trying to make someone believe something is true when it is not, which assumes an awareness that others have beliefs in the first place. *Homo sapiens* alone seem to have the capacity for flexible deception (doing it in many different contexts). Although some animals show one-off forms of deception, they don't seem to show generative or flexible deception (doing it in lots of different contexts) and may have simply learned a few algorithms (or rules).[49] This is an example of "Occam's razor": we shouldn't ascribe to an animal a complex psychological mechanism to explain its behavior if we can explain it more parsimoniously by reference to simpler psychological mechanisms. By the same token, the most parsimonious explanation of human flexible deception is in terms of a full theory of mind.

The capacity for flexible deception was a huge advantage in terms of natural selection. Not just mimicking being a stick when you're actually an insect, or mimicking being a rock when you're actually a fish, but being able to pretend *anything* you want, as we humans can. Just think about being able to scatter a carpet of leaves on a lattice of thin branches above a deep pit you've dug, pretending it's solid ground, and then waiting for your unsuspecting prey or a rival to walk into the trap, falling into the pit. *Homo sapiens* has this capacity to deceive flexibly, to plant false beliefs in the mind of their victim, and to vary this according to the context. One might wonder: What kind of evidence could we look for that Neanderthals, for example, could deceive? If they'd been able to lay a trap to kill another animal, or to use a stealth weapon at a distance, such as a dart or an arrow, this could have stood the test of time. The archaeological data currently show no evidence that Neanderthals had this capability.[50]

Second, a theory of mind allows for *flexible teaching*, because this involves keeping track of what another person knows or needs to know. Again, humans alone seem to engage in teaching. Evolutionary biologist Kevin Laland argues that other animals do show some evidence of teaching.[51] He cites the example of adult meerkats bringing disabled prey to their young offspring—such as a scorpion with the stinger removed—so they can learn how to kill them safely. This is certainly an intriguing example of a parental behavior that is likely to have adaptive benefits for both parent and child, one that may have been selected in evolution and is unlikely to be entirely encoded in the genome, as it involves flexible learning in both adult and child. However, one might question whether it qualifies as an example of teaching, as it does not necessarily require a theory of mind. Moreover, if it were a marker of a theory of mind, we would see a far wider range of examples of teaching in meerkats and other species, as well as evidence of the teaching strategies being *modified* by the teachers in response to what the pupils need to know.

And third, a theory of mind allows *flexible referential communication* because this involves understanding that your listener or your viewer knows that words (or paint marks on a cave wall) *refer* to something out there in the world.[52] One of the earliest signs of a theory of mind and of flexible referential communication is when human children as young as fourteen months use the pointing gesture, as if to say (without words), "Look at that!"[53] They are using their outstretched index finger to refer to something. Essentially, they are referring without words and establishing and sharing the topic of reference with another person, intending to influence that person's point of view. In humans, pointing is a good predictor of rate of language development, unsurprisingly, since pointing is a gestural equivalent of making a comment, and an early indicator of understanding that another person has a mind with which you can communicate. Despite claims to the contrary, I'm not convinced there are any persuasive examples of non-human animals using pointing in the wild.

And of course, referential communication enabled us to invent and understand drama and storytelling, where we establish a shared topic and describe characters who may have different points of view. It also allowed us to produce and appreciate humor, where we use words intending our listener not to take them literally, and to share information with others about objects or events beyond the here and now. Indeed, a suite of benefits flowed from having a theory of mind, such as conflict resolution, by being able to appreciate other points of view; the use of symbols, by intending an audience to understand our intention for one thing to represent something else; and new levels of social cooperation, by being able to form and communicate a plan with others for how to achieve a shared goal. Figure 2.7 shows how a theory of mind, part of the Empathy Circuit, was part of the cognitive revolution.

The absence of flexible deception, teaching, and referential communication in other animals is neatly explained by their not having a full theory of mind. Some argue that apes and monkeys and even rats have rudimentary levels of empathy, since they show some helping behavior toward members of

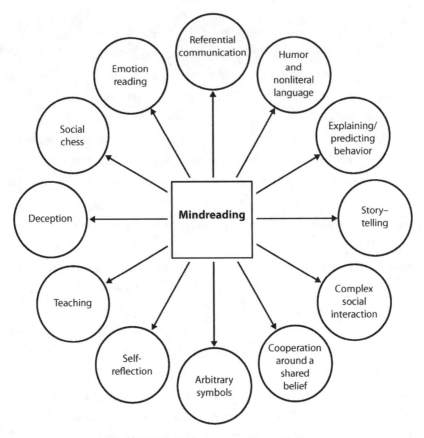

Figure 2.7. A theory of mind enabled a suite of behaviors unique to *Homo sapiens*.

their own species, or even toward members of a different species (dolphins coming to the aid of a human, for example).[54] But this could just mean they can recognize another animal is in distress.

Recognizing the distress call of another animal is very different from having a full-blown theory of mind or achieving what a four-year-old human child is capable of: understanding that another person or animal might have a *belief* that is different from your own. A human child as young as four years old can pass a "false-belief" test, the litmus test of a full-blown theory of mind.[55] In this test, developed by child psychologists Josef

Perner and Heinz Wimmer, a child watches a scene unfold between two dolls, Sally and Anne. In order to pass, the child has to keep track of their belief (what they know to be true), and what Sally (falsely) believes to be true.

A human child is able to recognize that, because Sally left the scene for a few moments, she won't know that, while she was gone, Anne moved an object (a marble) from one hiding place to another. When asked, "Where will Sally look for her marble?" the child will point to where Sally *thinks* the marble is, even though the child knows the marble was moved to the new hiding place. In doing so, the child passes the test. This is strong evidence that, even at this young age, a human child can imagine what someone else thinks. This little act of deception, which even a four-year-old human child can appreciate, lies at the heart of all good drama in theater, film, and novels, where, because a character may not know all the facts, different characters can have different beliefs about the same situation. Consider how the dramatic tension that builds in *Little Red Riding Hood* is because she *believes* the person in the bed is her grandmother when it is really the wolf. By contrast, there is no watertight evidence that understanding another animal can have a false belief is within the appreciation of non-human animals.

Some have argued that if you use gaze tracking, apes anticipate where someone will look if they hold a mistaken belief.[56] Personally, I'm not convinced that this is an index that apes have a theory of mind since, if they did, we'd expect the suite of behaviors discussed above. Yet others have argued that crows have a theory of mind because they wait to retrieve food when other animals are not looking, but this has been challenged on the grounds that the same behavior could arise for other reasons, such as stress.[57] There are claims that dolphins have a theory of mind because they seem to be taking into account what someone else knows, but critics argue that all they are doing is following cues about the direction in which a person is facing.[58] Interpreting the evidence from experiments with animals requires caution to avoid overinterpretation and anthropomorphizing.

|| || ||

So, the cognitive revolution 70,000 to 100,000 years ago encompassed the evolution of two new mechanisms in the brain, the Systemizing Mechanism and the Empathy Circuit. Both of these worked together to give rise to our remarkable human capacity for language, which is a mix of *if-and-then* rules (for example, in syntax) and theory of mind (such as keeping track of what your listener needs to know or may misunderstand). We will return to language later.

But most relevant for the new theory of human invention is the Systemizing Mechanism. Systemizing led to the invention of new tools and techniques in a dazzling array of activities: in music, clothes making, and art; in carpentry, architecture, and environmental management; in mathematics, science, and engineering; and even in law, philosophy, and ethics—all systems of *if-and-then* rules and logic.

When we step back and look at these two remarkable brain mechanisms that set modern humans along a path diverging from all other animal species, we discover remarkable *diversity* in the population. Most people systemize or empathize at average levels. But in some people, like Jonah and Al, their Systemizing Mechanism is tuned to the maximum—they are hyper-systemizers—while their Empathy Circuit is tuned super-low.

These hyper-systemizers struggle with even the simplest of everyday social tasks, like making and keeping relationships, yet they can easily spot patterns in nature or via experimenting that others simply miss. They have the potential to be inventors, even if they are bewildered by why other people ignore

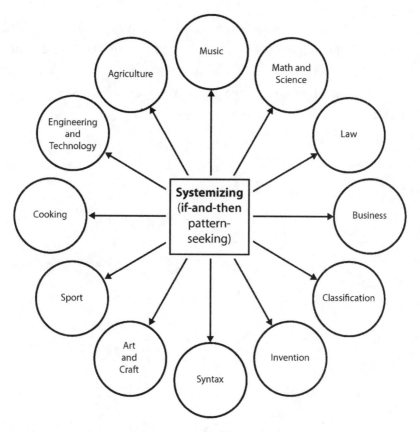

Figure 2.8. The Systemizing Mechanism enabled
a whole suite of behaviors unique to *Homo sapiens*.

them, or why others sometimes even reject or exploit them. In yet other people, hyper-empathizers, we see the opposite profile: super-high empathy (great at understanding people and being super-sensitive to what they might be thinking or how they might be feeling), but below-average systemizing. Of course, there are people in the population who are both hyper-systemizers (great at spotting patterns to figure out how things work) and hyper-empathizers, and we'll just look at the evidence for whether hyper-systemizers are more likely to struggle with cognitive empathy—whether there might be a trade-off between the two skills.

To understand this *neurodiversity* in the population we need to classify the different types of brain that exist. As it turns out, we can classify any population into just five types of brain, based on a person's levels of empathy and systemizing. Let's turn to look at which of the five brain types you have.

Chapter 3

Five Types of Brain

The UK Brain Types Study was an ambitious attempt to measure 600,000 people in terms of their empathy and their systemizing.[1] In this huge study, the first of its kind, we asked people to fill in brief versions of two questionnaires: the Systemizing Quotient (SQ) and the Empathy Quotient (EQ). (These are available in Appendix 1 if you want to find out how you score.)

The SQ asks about your level of interest in systems: these are as varied as map reading, music, knitting, grammatical rules, bicycle mechanics, cooking, medicine, genealogy, train time-tables, and public health.[2] All of these systems follow *if-and-then* rules. If you score high on the SQ, you are the kind of person who pays a lot of attention to detail, such as the small print in legal contracts, or the specifications of your computer or car engine. You might also be a collector of a niche category of objects, like coins, stamps, or butterflies, and you might like to keep an ordered (systematic) list of your top ten favorite songs or films (even if it's just in your head). Those who fall super-high on the systemizing bell curve are those who also have a strong drive to invent, and we can assume that, across human evolution, they always did.

In contrast, the EQ measures how easily you find it to imagine what another person might be thinking or feeling. As we discussed in the last chapter, cognitive empathy is the ability to imagine another mind, particularly what another person, animal, or any other kind of being (even a god) might believe, know, desire, perceive, or feel. This is the capacity that allows humans to engage in flexible deception, flexible communication, teaching, cooperation around shared beliefs, and even spirituality. These two measures, the EQ and the SQ, tap into the two revolutionary mechanisms in the evolution of the human brain.

The first thing we found in the UK Brain Types Study was that both empathy and systemizing fall on bell curves in the population. We are all located somewhere on each of these bell curves. A bell curve tells us that a trait lies on a continuum rather than being binary. A bell curve can also be a clue that a trait could be partly polygenic (meaning the trait could be partly influenced by hundreds or thousands of genes). Undoubtedly social learning also plays a role in empathy and systemizing, but the bell curves give us a clue that we should look for possible multiple genetic factors underlying these too.[3]

We then looked at five different brain types. The first group consisted of people who are equally good at both empathizing and systemizing. We called them *Type B*, for balanced. The second group comprised people who are high on empathy and low on systemizing. We called them *Type E*, because they are naturally drawn to empathize with people but less interested in systemizing to see how things work. In the third group were people who are the opposite: high on systemizing and low on empathy. We called them *Type S*, because they are naturally drawn to systemize but less interested in empathizing.

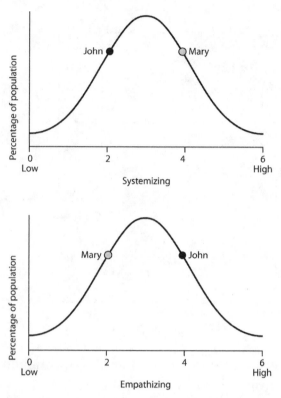

Figure 3.1. The systemizing and empathy bell curves. Mary, an engineer, is high on systemizing and lower on empathy. She is a hyper-systemizer. John, a therapist, has the reverse profile. He is high on empathy but lower on systemizing. He is a hyper-empathizer.

Finally, we looked at two extreme brain types: *Extreme Type E*, those who are super-high on empathy but below average on systemizing, and their polar opposite, *Extreme Type S*, those who are super-high on systemizing but below average on empathy. Jonah and Al, whom you met in chapter 1, are both Extreme Type S.

These five brain types are examples of *neurodiversity*—the varieties of brains we find in any school classroom or in any workplace, like the varieties of flowers or animals in nature.[4] None is better or worse than another; they are all just different,

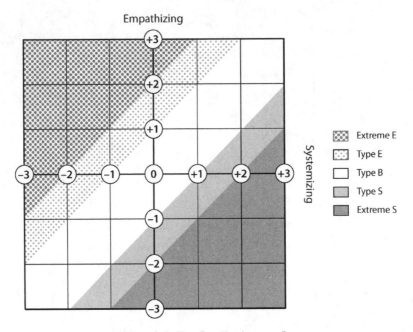

Figure 3.2. The five "brain types"

and each evolved to thrive in different environments. Which brain type you have depends on the tuning of the Empathy Circuit and the Systemizing Mechanism, which in turn determines where you are on the empathy and systemizing bell curves. To give you a feel of the different brain types, which are defined by the *difference* between your levels of empathy and your systemizing, let's explore them in a bit more detail.

TYPE E

These are individuals whose empathy is higher than their systemizing. About a third of the population is Type E, and this brain type is about twice as common in women (40 percent of women) as in men (24 percent of men). Those with Type E brains are very comfortable with people, chatting easily, seeing the dynamics in relationships, and easily able to tune into how others are feeling, and they are likely to gravitate toward jobs in the caring professions, such as counseling or charity work.

Compared to individuals who have Extreme Type E brains, they are somewhat less focused on empathy all the time—they make time for themselves as well as for others.

In terms of their systemizing, people with Type E brains can see new patterns when these are pointed out, but this doesn't come naturally to them. They can use technology when they need to but are not passionate about it. If shown a function in a device, they can follow the instructions given to them, but they would struggle to discover how to repeat the operation if they forgot it or if they had to work out how to do it for themselves. They use technology at a basic level, but do not master it. They enjoy getting to know one system or simple device if it makes life easier, but they still find more complex technology challenging. In contrast, they empathize intuitively.

TYPE B

Individuals who are Type B show no difference between their empathy and their systemizing, which is why they are called "balanced." They comprise roughly another third of the population and are as common among men (31 percent of all men) as among women (30 percent of all women). They are equally good, or equally challenged, at using empathy or systemizing. They are as good at communication as they are at using technology.

TYPE S

The mirror opposite of Type E, people who are Type S, are better at systemizing than at empathizing. They comprise roughly another third of the population. Type S is about twice as common in men as in women (40 percent of men and 26 percent of women). They are good at navigating their way through systems and seeing patterns in how systems operate. Without needing a manual, they can pick up a device and understand it, confidently and quickly, through trial and error. They are willing to "have a go," to experiment to find out how things work.

These are the people who are the first to jump in to volunteer to try to fix something when it breaks, in any situation—a classroom or the workplace, at a barbeque or on a camping trip. Type S individuals are also happy to teach others how to use technology. They may not be inventors of technology, but they are very useful to have around. They gravitate toward the exact sciences or the STEM fields, and to music, architecture, and other analytic fields (law, linguistics, accountancy, philosophy, and proofreading), as well as crafts, sport, nature (including gardening), and cooking. These are all endeavors where searching for *if-and-then* patterns is a main focus. They also gravitate toward special environments where they can analyze the effects of changing one variable at a time, for example, laboratories, mechanical workshops, music, art or film studios, business, kitchens, or gardens. Some even use their *if-and-then* thinking to construct fantastic novels or dramas with elaborate plots, or to construct stories or arguments in a legal case. In their everyday lives, they also prefer to do just one thing at a time. They like systems, so their lives are more orderly and routine. They dislike trying to follow a conversation when several people are talking at once. As regards their empathy, they can get by, but find empathizing an effort rather than something that feels either intuitive, easy, or fun.

EXTREME TYPE E

People with this brain type have very strong empathy but are below average on systemizing. People who are Extreme Type E are rare in the population, and this brain type is more common in women than in men (3 percent of women but only 1 percent of men). These individuals are hyper-empathizers. They empathize effortlessly and intuitively because their Empathy Circuit is tuned super-high. They are quick to anticipate what someone will feel or think, and what's best to say or not say in a conversation. They can chat effortlessly, even if there are multiple conversations going on at once. They quickly notice if someone is feeling awkward or upset, often before anyone else does, and are always thinking ahead as to what would be nice to do for another person. They stay in touch with their friends because other people's changing situations are always on their radar.

But in terms of their systemizing, they barely notice patterns. They leave other people to operate technology for them because their Systemizing Mechanism is tuned pretty low. These individuals, given a new gadget, might get it out of the box, perhaps learn one or two basic functions, but rarely experiment to explore its potential. They can notice easy patterns (like multiplication tables, or rhymes to remember how many days are in a given month), but they struggle to understand more complex patterns. And at school, they tend to avoid subjects like mathematics. Because their mind isn't looking for repeating patterns, it's easy for them to "switch gears" when things happen unexpectedly. It just doesn't bother them. Their natural way of thinking is to be empathizing non-stop.

EXTREME TYPE S

At the other extreme are individuals who, like Jonah and Al, are very strong at systemizing but below average on empathy. Like Extreme Type E, this brain type is also quite rare in the population. Extreme Type S is the mirror image of Extreme Type E in being twice as common in men (4 percent of men) as in women (2 percent of women). They are hyper-systemizers because their minds are wired to seek out patterns all the time, and they include autistic people.

Individuals like this can spot complex patterns such as: "*if* in a leap year the days in a given month fall on specific days of the week, *and* you look twenty-eight years ahead, *then* the days in a given month will fall on the same specific days of the week as twenty-eight years earlier." Or they may choose bell ringing (campanology) as a hobby, because of the mathematical patterns involved.

Spotting such regularities in a system may mean someone who is Extreme Type S can say immediately on which day of the week any date in the past or the future will fall. Such individuals are called "calendrical calculators," a form of "savantism." Savantism is when a person has one area of skill that is not only way above their other skills but is also way above most people in the general population. (Some estimates of savantism are one in a million in the general population but one in two hundred autistic people).[5] So, savantism is much more common among autistic people, and autistic people are more likely to be either Type S or Extreme Type S. Those who are Extreme Type S are also more likely to be autistic. People who are Extreme Type S are good at spotting deviations or inconsistencies in a system, at checking errors, and at fixing problems to make the system more efficient.

In terms of their social skills, because their empathy is below average, people who are Extreme Type S are often a bit too blunt in their communication, blurting out what they think or what's on their mind without necessarily considering the impact of their words, not sugaring the pill, and they may not see anything wrong with what they're saying or how they're saying it, defending it as factual. They struggle to make or keep friends, and their challenges with empathy may leave them vulnerable to exploitation by others—they just don't see the trap they are walking into. They are also at risk of feeling depressed after a long period of feeling socially excluded, despite their repeated attempts to try to fit in and to socialize with others. They may also mostly prefer their own company to that of others.

If you want to find out which brain type you have, just take the EQ and the SQ in appendix 1, which has a table to look up your brain type. You can also go online to find out at www.yourbraintype.com.

||||

It's interesting that most people in the population are not Type B—that is, equally good at both empathy and systemizing—when one would have expected this to be the optimal brain type. So why are only a third of people Type B?

In part this reflects how we have defined the brain types statistically to show us different percentiles in the population: Type B individuals are defined as those who fall between the 35th and the 65th percentile. But the fact that the majority of the population (two-thirds) are specialized to be either an empathizer (Type E) or a systemizer (Type S) may also be a clue that these brain types evolved under the pressure of natural selection, with some survival advantages to having a brain that was specialized. That would make sense if empathizers do well in one environment (the world of people, intuitively knowing what others may think or feel) and if systemizers do well in a different environment (the world of objects, figuring out how things work).

And why are only a small fraction of the population, about 3 percent, hyper-systemizers (Extreme Type S)? If, as I argue, being a hyper-systemizer is a prerequisite for the capacity for human invention, wouldn't you imagine that natural selection would have favored individuals with this brain type? Shouldn't they be far more numerous?

Again, a mundane statistical reason there are relatively few individuals with Extreme Type S brains is that we defined this brain type as those falling in the 2.5th percentile. But this may also reflect that there was negative selection pressure on individuals with this brain type. Perhaps as well as conferring remarkable advantages in spotting *if-and-then* patterns this brain type also carries certain disadvantages, namely, being challenged socially. This would be expected if the tuning of the Empathy Circuit and the Systemizing Mechanism is a *zero-sum game.*[6] That is, the higher one of these is tuned, the lower the other is tuned. We'll come back to look at evidence for this relationship between these two brain circuits.

Let's have a closer look at hyper-systemizers, who, according to my theory, are central to the story of human invention because they systemize non-stop. Over the last 70,000 to 100,000 years, we can assume that hyper-systemizers were the individuals who could make a new musical instrument (such as a flute), identify a new food (such as cultivating rice), develop a useful new skill (such as navigating by the stars), or create a new tool (an aqueduct, for instance). We can make this assumption because, when we look at modern inventors, many of them show the hyper-systemizing profile.

||||||

Do hyper-systemizers share the same kind of mind as autistic people? To answer this question, we went back to the UK Brain Types Study. Over 36,000 autistic people took part in this study, the largest ever psychological study of autism. We found that a disproportionate number of autistic people had a Type S or an Extreme Type S brain. Indeed, 62 percent of autistic males showed one of these two brain types, a higher percentage than the 44 percent of typical males with these profiles. And 50 percent of autistic females showed one of these two brain types, a proportion twice as high as that seen in typical females (27 percent). These results fit with the idea that the autistic mind and the hyper-systemizing mind have something in common.

As an aside, these results also fit with a different idea, that autistic people have a more "masculinized" profile, that is, they show the Type S or Extreme Type S brain types that are more common in typical males in the population and where their systemizing is higher than their empathy. You can see this shift very clearly in their D-scores, which are a measure of the *differences* between one's empathy and one's systemizing (see figure 3.3).

A further clue that autistic people and hyper-systemizers share a similar type of mind comes from a study in which we gave autistic young teenagers a mechanical reasoning test where they had to figure out how a novel system worked. The test was originally designed to identify adults with the potential to study engineering. Autistic teenagers consistently outperformed typical teenagers on these tests.[7]

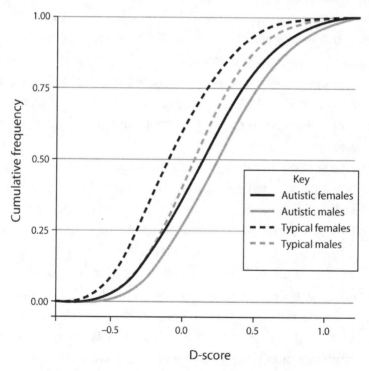

Figure 3.3. Your D-score is the difference between how much higher your systemizing score is relative to your empathy score. Typical males (dashed gray line) have a higher D-score than do typical females (dashed black line). Autistic females (solid black line) and autistic males (solid gray line) have extreme D-scores.

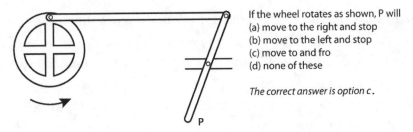

If the wheel rotates as shown, P will
(a) move to the right and stop
(b) move to the left and stop
(c) move to and fro
(d) none of these

The correct answer is option c.

Figure 3.4. An example from the mechanical reasoning test

And there are more clues that this is the case. Psychologist Laurent Mottron found that autistic people were 40 percent faster than typical individuals at pattern detection using a nonverbal visual intelligence test, and that they showed more activity in visual areas of the brain while achieving this. And a study conducted in Silicon Valley in 2013 found that more autistic students were studying a STEM subject at university, compared to any other disability group, again suggesting that autistic people have hyper-systemizing minds.[8] This was demonstrated back in 1972 by psychologist Uta Frith in her finding of superior pattern production by autistic children, even in those with a learning disability.[9]

Here's a second question: Do systemizers (people with a Type S brain) and hyper-systemizers (those who are Extreme Type S) have more autistic traits? In the Cambridge University Autistic Traits Study, we asked over 1,000 students to take the Autism Spectrum Quotient (AQ).[10] (Everyone has some autistic traits on the AQ, but some of us have more traits than others, and autistic people have very high scores on the AQ. If you want to see how many autistic traits you have, you can take the AQ in appendix 2.) We found that students in STEM disciplines had more autistic traits than students in the humanities. And we found that math students were more likely to have a diagnosis of autism, compared to students in the humanities.[11]

To extend this, we conducted the Big AQ Study, in which half a million people worldwide took the test—at the time, the largest study of autistic traits on the planet. Sure enough, we found that those working in STEM scored higher on the AQ compared to those working in non-STEM occupations. And we confirmed this in the UK Brain Types Study, this time with 600,000 people.[12] Not only were those in STEM more likely to be Type S or Extreme Type S, compared to those in non-STEM occupations, but those in STEM occupations on average also had a higher AQ. In the UK Brain Types Study, we also found that people who were Type S or Extreme Type S had a higher score on the AQ.

This nailed it for us. Those with autism and those who are strong systemizers have similar minds.

▌▏▌▏▌▏

Let's return to the idea that systemizing and empathizing can be a zero-sum game: the more you have of one capacity, the less you have of the other. If this is the case, we should expect to see a trade-off, a negative correlation between empathy and systemizing: the better you are at one, the worse you are at the other. The results from the UK Brain Types Study did show a small trade-off.[13] This suggests that, while these two dimensions are largely independent of each other, they may also share a common biological factor. But what could that be?

The biological factor might be expected to be a molecule that differs in quantity between the two sexes. That's because we found three times as many females as males among those who were Extreme Type E, and twice as many males as females among those who were Extreme Type S. One such candidate biological factor is the amount of the hormone testosterone that a fetus's brain is exposed to in the womb.[14] Male fetuses produce at least twice as much of this hormone as female fetuses do during prenatal life, when the brain is developing, and animal research shows how this hormone changes (or "masculinizes") the brain.

Talking about a more masculine brain, in humans, is controversial; some people want to deny that there are any sex differences on average in the human brain.[15] But there is now no longer any question that there are some key sex differences on average in human brain structure, and these emerge when large-scale brain scanning is conducted. One convincing set of data to illustrate this comes from the UK Biobank (see figure 3.5).

There are also sex differences on average in the number of neurons that males and females have, as counted in post-mortem studies of the brain. (Females on average have 19.3 billion neurons, while males on average have 22.8 billion. See figure 3.6.) And several of these average sex differences have been demonstrated at birth, suggesting that, whatever the role of culture later, a prenatal biological factor is also involved.[16]

So, could one reason why there are more females who have Type E brains and more males who have Type S brains be that males are exposed to higher levels of testosterone in the womb? We set out to test if prenatal testosterone shapes brain type, through the UK Prenatal Testosterone Study.[17] This study followed six hundred babies in the United Kingdom from the womb into their teens. We measured their testosterone levels before they were born, to see if this predicted their later empathy and systemizing levels. We were able to make these measurements because all of the babies had mothers who had opted, for clinical reasons, to have an amniocentesis test. This is where doctors use a long needle to extract some of the amniotic fluid surrounding the fetus, to check for fetal anomalies. This gave us an opportunity to measure levels of prenatal testosterone in the amniotic fluid.

Prenatal testosterone (and prenatal estrogen, the hormone that testosterone is converted into) has long been proposed as one driver of the differences seen on average in the brains of males and females, based on more than half a century of animal

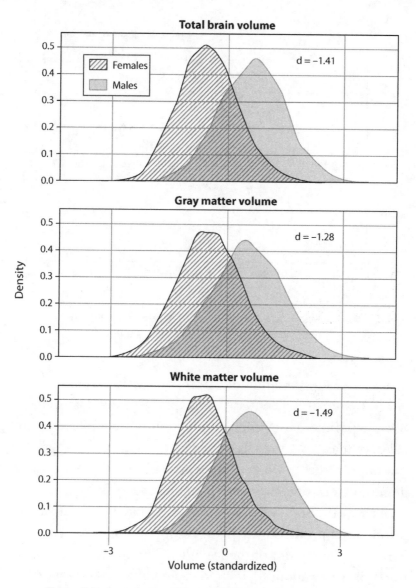

Figure 3.5. Average sex differences in gray matter, white matter, and total brain volume, in the adult brain of over 5,000 individuals. Hatched gray = females; solid gray = males. All of these average sex differences are significant after controlling for total brain volume, total surface area, average cortical thickness, and height.

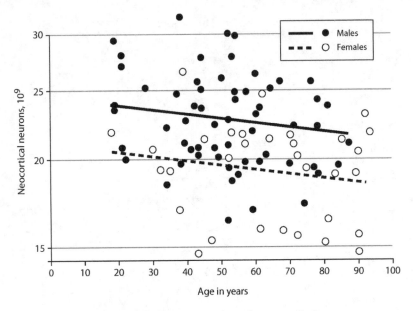

Figure 3.6. Average number of nerve cells in
males and females in the cortex of the brain

research where the amount of prenatal testosterone or estro-gen the animal is exposed to is experimentally manipulated. Such experimental manipulations would of course be unethical in humans, but the UK Prenatal Testosterone Study gave us the opportunity to observe the correlations between prenatal amniotic hormone levels and the child's later behavior.

As hypothesized, we discovered that a baby's prenatal testosterone level predicts if they will end up being more Type E or Type S, independent of their gender. The mothers completed questionnaires about their child at age four, and we found that the higher the baby's prenatal testosterone, the higher the child scored on the SQ and the lower they scored on the EQ. We also found that the higher the baby's prenatal testosterone, the higher they scored when we followed them up at age eight on an attention-to-detail (systemizing) test and the lower they scored on the "Reading the Mind in the Eyes" (empathy) test.[18]

This was remarkable: testosterone levels in the womb were associated with *both* systemizing and empathizing, but in opposite directions. Prenatal testosterone was a biological factor that could partly explain why brains ended up leaning either toward Type E or toward Type S. And when we examined MRI scans of these children's brains, we found that prenatal testosterone levels were indeed positively correlated with some brain regions (like the superior temporal sulcus, a region tuned to detect where someone is looking) and negatively correlated with others (like the planum temporale, a region involved in language).[19] How much prenatal testosterone a child was exposed to in the womb is one of the factors associated with how fast they develop speech, and it is well established that girls on average develop language faster than do boys.[20]

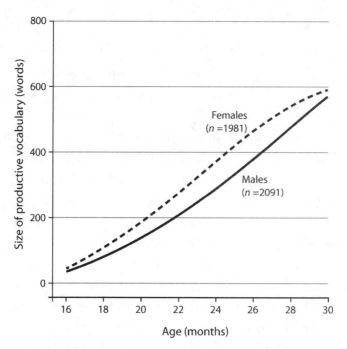

Figure 3.7. Average size of vocabulary produced by boys and girls between sixteen and thirty months of age

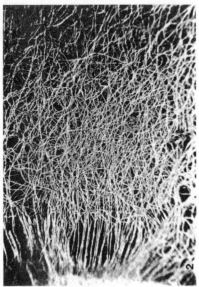

Figure 3.8. The effects of adding estrogen on neurons and their connections in the mouse brain. The brain region is called the sexually dimorphic nucleus of the preoptic area (SDN-POA). *Left*: no estrogen added. *Right*: estrogen added.

To give you an idea of what sex hormones like testosterone and estrogen can do to the developing brain, look what happens if you add extra estrogen to a key region in the mouse brain (see figure 3.8).

Given that some hyper-systemizers, like Jonah, are autistic, does how much testosterone a baby is exposed to in the womb predict how many autistic traits they will have as they grow up? We found that the higher a child's prenatal testosterone, the more autistic traits they had.[21] This was seen in the mother's answers on a questionnaire measuring autistic traits in her toddler at eighteen months old, and again when her child was four years old. So how much you empathize, or how much you systemize, or how many autistic traits you have is influenced by how much prenatal testosterone you were exposed to in the womb.

To put this to the ultimate test, we conducted a study to find out if hyper-systemizers, like Jonah and other autistic children who struggle with empathy but who can be excellent systemizers,

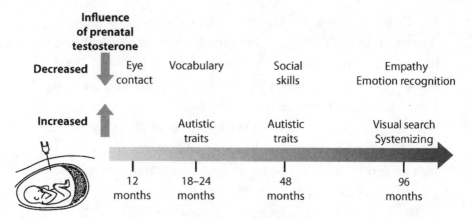

Figure 3.9. Higher prenatal exposure to testosterone is associated with both better systemizing and reduced empathy.

were exposed to elevated prenatal testosterone. We investigated this in the Danish Prenatal Testosterone Study, using the Biobank in Copenhagen, where obstetrician Bent Norgaard-Pedersen invited us to analyze the amniotic fluid from 20,000 samples he had carefully stored in his deep freezer since the mid-1990s. We looked at samples from pregnancies where the baby later received a diagnosis of autism and compared them to pregnancies where the baby developed typically.

Sure enough, autistic children on average had higher levels of prenatal testosterone than typical kids. Interestingly, this also had knock-on effects for their levels of prenatal estrogens, which were also elevated on average in those who were later diagnosed as autistic, as would be expected, since testosterone is converted into estrogen.[22] So, we had evidence that a biological factor—prenatal testosterone and estrogen—influences how much of an empathizer or a systemizer you are, and linking hyper-systemizers and autistic people at the hormonal level.

So much for prenatal hormones changing your brain, but can your genes change your brain too?

Many people would shudder at the idea that genes have any-thing to do with traits like systemizing or empathy, wanting to believe that all you need are the right learning opportunities and experiences. And they are partly right: learning and experience do count for something. But to claim that empathy and systemizing evolved, these psychological abilities must have had at least a partly genetic basis. To discover if genes play *any* part in empathy and systemizing we launched the ambitious Genetics of Empathizing and Systemizing Study.

Working with the personal genomics company 23andMe, we asked their customers—who consent to their genetic data being shared anonymously with researchers—to take two empathy tests so we could look for genetic associations with individual differences in scores on the empathy tests. Eighty-eight thousand people who had provided their DNA took one of our empathy tests, the "Reading the Mind in the Eyes" test, or the "Eyes" test for short.[23] Of these, 46,000 also took the other empathy test, the Empathy Quotient (EQ), and 50,000 took the Systemizing Quotient (SQ).

On the Eyes test, volunteers are shown photos of the eye region of the face of different actors, each with a different emotional expression, and are asked to choose which of four words best describes what the actor is thinking or feeling. The test assesses how well you can imagine the mind of another person. We then carried out an analysis across the genome. We were excited to discover that scores on this test were heritable (the heritability among twins was 28 percent).[24] This was sufficient to prove that empathy or theory of mind is partly under genetic control. And what's more, we found that a sequence in one gene on chromosome 3 is associated with how well women perform on this test. So, while how much empathy you have is of course in part the result of your social experience, genes also play a role.

To be doubly sure about this conclusion, we looked at the DNA of the 46,000 individuals who had taken the EQ. Again, we

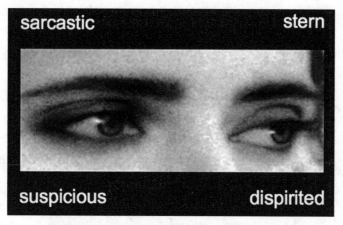

Figure 3.10. An item from the "Reading the Mind in the Eyes" test. Which of these four words best describes what the person in the photo is thinking or feeling? The correct answer here is "dispirited"—she's a little bit sad.

carried out an analysis across the genome. Once again, we found that a common genetic variant is associated with how they scored on the EQ, and that some of the variation on the EQ could be explained by common genetic variants.[25] Finding a genetic association with scores on empathy tests means that empathy evolved and could have been subject to selection pressure. (Selection pressure is where some individuals have a better chance of surviving to pass on their genes, giving rise to evolution.)

We then turned to systemizing to find out if that is also partly genetic. This time we looked at the 50,000 people who had been genotyped and who had taken the Systemizing Quotient. Excitingly, we found that some of the variation on the SQ could be explained by common genetic variants, and that common variants in three genes were associated with systemizing, independent of other factors.[26] So while the strength of your interest in systemizing is of course influenced by environmental factors (such as whether you had a good science teacher, or a parent who inspired you to observe nature, encouraged you to experiment with fixing things, or provided mechanical toys for you), we had proven that how much you systemize is also partly in your DNA—and therefore could have been positively selected in human evolution.

Another way to test if a trait is partly genetic is to study twins: if identical twins are more similar than are non-identical twins in a certain trait, this is a clue that genes play a role. One measure of systemizing that has been studied in twins is what Nobel Prize–winning psychologist Herbert Simon calls "satisficing."[27] Most of us are "satisficers": we use the first thing to hand that works well enough to solve a problem. We say to ourselves, "This is good enough," and we aren't perfectionist, just wanting to get the job done speedily. We cut corners. In contrast, strong systemizers (sometimes called "maximizers") search for the *optimal* solution to each problem that arises, even if this takes forever.

Imagine if you had to do some sewing. Most of us will just rifle through the sewing box at home, looking for a needle that will do the job, and use that. That's satisficing. But some people will know that the perfect needle for the particular kind of material is of a particular length, thickness, and eye size. They will keep searching in the sewing box until they find the exact needle they need, even if it takes hours. That's maximizing. Maximizers are perfectionist because they want to optimize the system, in this case sewing. So they are hyper-systemizers. A study of twins scored people on a dimension from satisficers through to maximizers. Identical twins were more alike on this dimension than were non-identical twins. So where you fall on the systemizing bell curve is partly genetic.

This fits with the results of the Genetics of Empathizing and Systemizing Study, where we found the common genetic variants associated with systemizing and empathy were independent of one another. This proves that the Empathy Circuit and the Systemizing Mechanism are under separate genetic influence.

What about the link between hyper-systemizing and autism? We saw that this link has a hormonal basis in the womb, but is this link also partly genetic? We know autism is partly genetic, for three reasons: First, while only 1 to 2 percent of the population has a formal diagnosis of autism, a younger child in a family with an older sibling diagnosed as autistic has a 10 to 20 percent likelihood of also being autistic.[28] Second, where one of a pair of twins is autistic, the other twin is more likely to be autistic too if he or she is an identical, rather than a non-identical, twin. Finally, there are more than one hundred rare genetic variants or mutations that have been linked to autism.[29]

As part of the Autism and Math Study, we asked math students at Cambridge University (who can be assumed to be strong systemizers), "Do you have a sibling who is autistic?" We found that the autism rate among the siblings of math students was higher than the autism rate among the siblings of students in the humanities.[30] This hints at a shared genetic basis for autism and systemizing, since siblings on average share 50 percent of their genes.

Rare genetic mutations have only been found in fewer than 5 percent of autistic people, so the genetic basis of autism in the other 95 percent of autistic people is likely to lie in common variants that we all carry but that occur in specific combinations in autism. We tested if the known common genetic variants for autism overlap with those associated with systemizing. Among the 50,000 people who took the Systemizing Quotient, we found that 26 percent of the common genetic variants associated with a high score on the SQ correlated positively with the common genetic variants associated with autism. This was enough to prove that hyper-systemizing and autism share a genetic source. Expressed differently, **some of the genes for hyper-systemizing and some of the genes for autism are the very same genes**.

▌▌▌▌▌

If you are someone who scores extremely high on the SQ and extremely low on the EQ, and so have an Extreme Type S brain type—that is, if you're a hyper-systemizer—this means you think differently compared to most people. That's because your mind, in common with most autistic people, has a *different operating system.* Less focused on people and more focused on things and on patterns, your operating system struggles to function in some environments but in others it may confer what autistic climate activist Greta Thunberg calls "superpowers": a talent at spotting *if-and-then* patterns, just like Jonah did as a child.[31]

Chapter 4

The Mind of an Inventor

Over the decades I have watched as Jonah grew into a wonderful man with remarkable talents.

ˌLike a modern-day Linnaeus, Jonah also loves systemizing the world of plants. Wherever he goes, his eyes are constantly noting the plants around him. Jonah told me he is checking each plant against what he calls his "mental spreadsheet," which is organized into rows and columns. As he describes it to me, each row (the *if*) might be a plant characteristic (the shape of a leaf, the color of a petal), each column (the *and*) might be an environmental variable (a preferred soil, the season when it blooms, its geographical location), and where these intersect (the *then*) might be the name of a specific plant. *If-and-then* reasoning. In his mind's eye, all Jonah has to do is read across from the plant name to identify what is special and unique about each plant.

Jonah started classifying types of leaves in the playground at age six, as you'll recall, and today his knowledge of plants is encyclopedic. His mental spreadsheet allows him to understand how different plants are related to each other, and all these patterns conform to the *if-and-then* algorithm. In this way, he does what many autistic people do: he records information systematically. Today he is obsessed with collecting information for every species of tree in the world—all 60,000-plus species.[1]

His memory has the hallmark of hypermnesia (the opposite of amnesia)—it seems to know no limits.[2] There are only a handful of documented individuals with hypermnesia, adults who can remember every day of their lives since at least age fourteen. With Jonah, his memory is for factual information about objects, and specifically plants.[3] His recall is also extremely rapid. His family describe him as being able to "read" nature. Sometimes they demonstrate this to others by pointing to a random plant and seeing how many facts about it he can rattle off. But Jonah isn't interested in showing off. He is simply driven to systemize plants accurately and comprehensively, interested in *facts*, *patterns*, and *truth*. For him, as for many autistic people, these three words all refer to the exact same thing.

Jonah's other passion as an adult is car engines. As a car drives by, he can tell from the sound alone if the engine is developing a fault and can diagnose which component needs replacing. The sound of each component in the engine is the *if*, tweaking it is the *and*, the car's efficiency is the *then*—again, he is constantly seeking *if-and-then* patterns. As with plants, he describes looking up these patterns in his mental spreadsheet. Jonah loves nothing better than being asked by someone to tune their car. Then he can lock himself away from the world, sometimes for days, undistracted by people, and concentrate 100 percent on the task, until he has checked, double-checked, and triple-checked every aspect of the car's engine and the engine is performing optimally. Sometimes he bluntly tells someone their car will develop a specific problem, which can cause anxiety in the listener. Their feelings are not first and foremost in his mind, though; he is simply compelled to tell the truth. His family has learned to heed his predictions, because they always prove to be right.

But despite his talents, Jonah's skills are underused. He has written more than four hundred unsuccessful job applications, with his parents' support. Although they encourage him to remain positive, he can see the facts for themselves and often feels depressed and hopeless. Being unemployed and living with his parents at age thirty-two makes him feel like he has not been accepted by society. On two occasions his depression got so bad that he felt like ending his life, and he has made two attempts to do so. I asked him:

"Did you want to die?"

He nodded, and when I asked him why, he simply replied:

"Nobody wants me. I don't belong in this world."

I acknowledged his loneliness, aware that over 80 percent of young autistic adults still live with their parents.[4] I asked him:

"What would make a difference, what would make life feel worth living?"

Without looking up, he replied:

"A job. To make me feel valued—to give me dignity. Why won't anyone give me the chance to prove I can contribute, to make me feel included in society, and give me a wage, so I can have independence from my parents?"

I nodded, feeling so sad for him and the millions of other autistic people who languish unemployed when they could be doing something meaningful for themselves while helping their employer and society.[5] He added:

> "I want what every human being has the right to: to have the basic financial means to make decisions about how to live. Unemployment is killing me, and so many people like me."

His analysis was totally accurate. We surveyed four hundred autistic adults like Jonah who had attended our clinic in Cambridge, and tragically we found that two-thirds of them had felt suicidal and one-third of them had actually attempted suicide.[6] What more of a wake-up call does society need that autistic people are struggling and desperately vulnerable? Jonah's life illustrates how huge is the waste of talent among these autistic hyper-systemizers, and how unemployment adds to their suffering from exclusion.

Unemployment is just one of the challenges Jonah faces. He often says he feels totally lost in the social world. He would like to have close friends and an intimate relationship, but these have never materialized for him. He finds conversation confusing, not knowing what to talk about, when it is his turn to speak, or what his conversational partner is expecting. These challenges reflect his difficulties with "cognitive empathy."[7] So he often avoids company and avoids the stress that conversation brings. When he tries to chat, he worries he is getting it all wrong. Despite his remarkable ability to systemize the world of plants or car engines, he just can't figure out how to keep a conversation going.

He says people either ignore him in a group or speak over or for him, because he is slower than others to reply. He hates the telephone, because he doesn't know what to say and finds the silences painful. Some people have told him his voice sounds monotonous, and that he speaks too loudly, but he doesn't know how to make his voice sound any different. He readily admits that he can't imagine what other people hear, or how he might come across to them. To do so would involve imagining what someone else thinks or feels, which he finds totally mysterious and beyond him.

When Jonah is in a social group, he often feels that everyone else understands a joke—they all seem to laugh at the same time—and he is left wondering what he missed. This difficulty in understanding humor is evident even in autistic toddlers, compared to typical toddlers, who love to joke around and who can switch easily between serious and playful communication.[8] People tell Jonah that understanding humor is all about reading between the lines, but all he can deal with is factual information, not implicit meanings. He notices other people exchanging glances, shrugging, or raising an eyebrow, but he has no clue how to interpret such body language. He says it's as though everyone else is speaking a silent private language that he doesn't understand. It makes him feel as if he comes from another planet, that he is watching a complex species from the sidelines of a game that he cannot join, and from which he even feels he is actively excluded.[9]

His challenges with cognitive empathy are typical of many autistic people.[10] But while he struggles with this, people who know Jonah describe him as very caring. For example, if he hears that someone is unwell, he tries to think what he could do for them, to help them. If he hears that someone has been treated unfairly or is suffering, it upsets him, and he wants to do something about it. So as with many autistic people, his "affective" empathy is intact.[11] In this sense, an autistic person is the mirror image of a psychopath, whose cognitive empathy is often highly practiced (to exploit others) while their affective empathy is blunted. Psychopaths, unlike autistic people, just don't care how others feel.

Jonah feels bitter about his childhood. Having been relentlessly bullied, both physically and verbally, has left a mark on his self-confidence, and he attributes his depression in adulthood to this. He thinks if other children had just left him alone, he could have been happy then, and would be happy now. But instead, they teased and mocked him mercilessly as a child, made him feel like a failure, and he still feels like that now.

Jonah is one example of an autistic hyper-systemizer who has struggled. But not all autistic people struggle: Daniel Tammet is another hyper-systemizer who came to our clinic and whose life has turned out remarkably.[12] I gave him the diagnosis of Asperger syndrome, a term that was given at that time to a subgroup of individuals on the autism spectrum who had at least average language and intelligence.[13] Like Jonah, Daniel has a remarkable mind. For example, he memorized the number *pi* (π) to 22,514 decimal places, and after reciting it (it took him five hours and was invigilated), he was awarded the title of European champion in this memory competition. I asked him:

"Why did you take on this challenge?"

He smiled at me, and gently replied:

"Memorizing a sequence like *pi* is comforting and reassuring, because *pi* is always the same: 100 percent predictable. It's the exact ratio of a circle's circumference to its diameter. Isn't that beautiful? Just as some autistic children line up their colored bricks or toy cars in long sequences, which follow a logical order, for me numbers give me peace and pleasure as they always fit together in the same, reliable pattern."

I nodded, admiring his remarkable mind, and asked:

"What do numbers mean to you?"

He looked up and said:

"As a child, I was stressed by people, because there's no pattern to their behavior—they never do the same thing twice. So I made friends with numbers, rather than making friends with my classmates. In my mind, I break long number strings down into their component parts, and then spot the patterns. I can multiply three-digit numbers together, sometimes faster than a hand calculator, reassembling numbers from these basic units."

Daniel, who has learned ten languages, also analyzes human language in the same way, rapidly spotting patterns in grammar and collecting tens of thousands of words, much as Jonah collects plant names. A television crew tested Daniel's language learning ability by taking him to Iceland, without any prior warning, because they knew he had no knowledge of Icelandic. After spending just one week there, he was interviewed in Icelandic on Icelandic television and performed well. But alongside his talent in memorization, numerical calculation, and languages, Daniel had all the classic signs of autism in his childhood. Like Jonah, he kept to himself at school and didn't make eye contact until he was twelve years old because he didn't realize it was important to do so.

Talking with Daniel, I realized that when you see an autistic child spinning the wheel of a toy car, held close to their eye, they are likely spotting unchanging patterns. I've seen this in autistic children with additional learning disability or below-average IQ as well as in autistic children with average intelligence or above. Just as a wheel goes around and around, repeating patterns, so *pi* doesn't change. Some autistic people are drawn to more concrete repeating patterns, like the spinning car wheel or a spinning fan or a spinning washing machine, while others are drawn to more abstract repeating patterns, like *pi*. Whether their interest is more concrete or more abstract may be influenced by their IQ, but autistic people irrespective of IQ are tuned to look for *if-and-then* patterns so they can discover *constants*, just as a scientist or a mathematician does. Not for no reason is *pi* called a mathematical constant, meaning it *always* applies, to *every* circle. I realized that what fascinated Daniel about *pi* is the very same thing that fascinated the Sicilian mathematician Archimedes more than two thousand years ago, in 250 BC.

||||||

The positive side of hyper-systemizing is that it confers an advantage in the capacity to see *if-and-then* patterns, to analyze and fix systems, and to invent new systems. But there can be a downside: some hyper-systemizers get into trouble because their hyper-systemizing leads them to pursue their pattern-seeking obsessively, with a kind of tunnel vision. Some hyper-systemizers become blind to the risks of their behavior, for themselves or for others.

Take Lauri Love, who was a British student of electrical engineering and who for five years faced extradition to the United States on charges of hacking into America's military computer network to steal data. I met him at the request of his lawyer, to see if he was autistic. It was clear that Lauri warranted this diagnosis. As I probed deeper, I realized that he was not a criminal in the sense of being motivated to profit from others. Rather, his motivation for this alleged crime was ethical hacking, in what he believed was in the public interest, albeit in an obsessive manner. Here's what he said:

"We hackers are trying to improve the web and digital security. Take TalkTalk, which was hacked by a fifteen-year-old, and in doing so helped that company to realize its security was broken. Hackers can use technology for good, but technology can also be used for bad. For example, GCHQ and the NSA have the power of surveillance of everything. They can hack into your SIM cards to steal private keys."

Lauri showed me his laptop. On the web browser, he had tabs open for hundreds of websites. I was astonished, as he could name all the websites from memory and recall the information he had read on each of them. Just like Jonah, with his vast memory for plant names, or Daniel, with his vast memory for numbers, Lauri's vast memory was for websites. He described his interest in computing as a compulsion. But in his single-minded pursuit of information, he had never considered the possibility that he might be branded as a "terrorist" or face life imprisonment for his alleged activities, because hurting people was the very opposite of his values. He told me bluntly:

"If I am extradited to face trial in the United States, I will kill myself, because I would never survive the brutality of an American prison."

The thought of being jailed was overwhelming for Lauri because autistic people are hyper-sensitive to sounds, lights, unfamiliar people, and unexpected change.[14] This doesn't begin to convey the intolerable suffering they may experience, and he knew prisons could also be violent places, totally unsuitable for a vulnerable person. When Lauri was finally told that the UK government was not going to hand him over to US authorities to stand trial, and that he would not be charged under UK law, he was immeasurably relieved. I was delighted that the UK courts had the good sense to realize this autistic young man, a hyper-systemizer, would be better off staying with his family than being thrown into jail, and would be unlikely to reoffend.[15]

||||

Jonah, Daniel, and Lauri—all are autistic young men with hyper-systemizing minds who you could argue show the mark of "genius." And I've met many similar autistic women, particularly here in Cambridge University, who are super-talented scientists. Genius is sometimes defined as looking at the same information that others have looked at before and either noticing a pattern that others have missed or coming up with a new pattern that constitutes an invention.[16]

It's not that all autistic people are geniuses. Autism is a wide spectrum that includes those with learning disability. But what we do know is that a disproportionate number of autistic people are hyper-systemizers, and that hyper-systemizers, with their talent at spotting novel patterns, therefore have potential to be inventors. Recall Al, with his moonlight experimenting. How did his life turn out?

||||

Thomas "Al" Edison couldn't stop systemizing.[17] But he also struggled to understand people, and posthumously he has been described as autistic. Had he lived today, he might not have sought or needed a formal diagnosis, but he clearly had a high number of autistic traits. He was obsessively experimenting during his childhood in the basement of his house and continued experimenting into his teens—so much so that, when he was supposed to be selling newspapers on the train, he was actually testing chemical reactions in the baggage car. He was so blind to everything but his experiments that he didn't think of the possible risks to himself or other people. On one train journey, the chemicals exploded, causing a fire, and he was lucky not to lose his job.

When Edison was in his twenties, trying to make his way as an inventor, he fell deep into debt. One day he begged a cup of tea from a street vendor, and while he was drinking it, he noticed that the manager of a local company had a broken stock-ticker (a device for continuously reporting the latest prices on the stock exchange). He couldn't help himself—he went over and fixed it. The grateful manager hired Edison on the spot. This was a fortuitous turn in his prospects.

For the next twenty years, Edison's inventiveness was unstoppable. By age twenty-nine, he had invented the carbon transmitter, which made Alexander Graham Bell's telephone usable. By thirty-two, he had invented the first commercially feasible lightbulb. By thirty-six, he had built the first economically viable system of centrally generating and distributing electric light, heat, and power. By forty-three, he had invented the Vitascope, an early film projector, which led to the first silent motion pictures. Edison also invented the first practical Dictaphone, mimeograph, and storage battery. He was a tireless hyper-systemizer. He is famously alleged to have said about his relentless experimenting:

> "I have not failed; I have simply found ten thousand ways that won't work."[18]

This remark perfectly reflects the need to try every variable in a system and to keep track of the effects of these systematic changes, which is at the heart of the Systemizing Mechanism.

But despite the obvious talents of his hyper-systemizing mind, Edison's obsession with work and his poor social skills continued to cause difficulties. He forged few close relationships. He had married his first wife Mary in 1871, when he was twenty-four and she was just sixteen. Together they had three children, the first two of whom were nicknamed Dot and Dash (reflecting his childhood interest in Morse code). Mary died at age twenty-nine, and at the age of thirty-nine, he married Mina, who was just twenty years old—virtually half his age—and had three more children with her.

Throughout his marriages, Edison

worked 18 hours a day or longer, pushing ahead for weeks on end, ignoring family obligations, taking meals at his desk, refusing to pause for sleep or showers. He disliked bathing and usually smelled powerfully of sweat and chemical solvents. When fatigue overcame him, he would crawl under his table for a catnap or stretch out on any available surface. Eventually his wife placed a bed in the library of his...laboratory.

Edison's method of inventing, and evidence of his hyper-systemizing mind, was to use

a dogged, systematic exploratory process. He tried to isolate useful materials—his stock room was replete with everything from copper wire to horses' hoofs and rams' horns, until he happened upon a patentable, and marketable, combination.

You can imagine walking into this room where he had saved everything in case it might be useful for something at some point. As he searched for solutions to problems, he simply looked around to try this component or that one, continuously experimenting, the seeming piles of junk a veritable Aladdin's cave of miscellaneous and unusual objects that served themselves up to his hyper-systemizing mind. "*If* I measure X, *and* substitute A with B, *then* X increases. But *if* I measure X, *and* substitute A with C, *then* X decreases." Simple experimenting, which Edison had been doing since childhood, and which *Homo sapiens* has been doing for 70,000 to 100,000 years. Evolution has come up with only one kind of algorithm in the brain to enable invention, and in Edison's brain this algorithm was tuned to the maximum.

He became embroiled in a long argument with Nikola Tesla about the use of electricity. Tesla was a rival engineer (also described as autistic, because he was hyper-sensitive to light and sound, obsessed with the number 3, and socially difficult).[19] Their polarized stances made it impossible for them to collaborate, even though doing so could well have paved the way for other remarkable inventions. Their argument may have been a reflection of limited empathy: each believed that there could be only one correct view—his own—and that the other must be wrong. And their limited empathy may have left them unmotivated to try to reconcile different perspectives or to acknowledge equally valid points of view.

There are several indicators that Edison's empathy was quite limited. For example, some of his experiments led to inventions that other people simply didn't want, which, in his solitary and obsessive state, he simply didn't foresee. One example is his invention of the talking "Edison doll," which children simply didn't like. He hadn't bothered to check with any actual children to see if this was something they would enjoy. To hear the doll recite a nursery rhyme, you had to crank a handle, and to hear it recite a different nursery rhyme, you had to open the doll and replace the small phonograph record with another one, a fiddly procedure.

While any parent could have told him that most children would quickly get bored hearing the same rhyme again and again, Edison hadn't checked what they might like or dislike, their feelings, and therefore hadn't anticipated how they would react. And he also hadn't appreciated that children wouldn't have the patience to figure out how to change the record. Nor had he thought through that the voice of the doll, which was high-pitched and monotonous, would be unattractive, even scary. These were all signs that Edison didn't put himself into other people's shoes—that he had reduced cognitive empathy. Unsurprisingly, the doll was a total commercial failure. Of the 2,500 dolls distributed to stores, no more than 500 were sold, and production of them ceased after just a few weeks.

A second example of Edison's limited empathy was his design for a concrete home fitted with concrete furniture that could be mass-produced using an intricate mold. He tried to sell the idea for seven years, even offering it for free to builders, until he was forced to accept that what he believed was a brilliant idea was a non-starter.

But occasionally Edison's inventions met a pressing public need, as when he came up with the formula for a reliable lightbulb. His ceaseless drive to systemize thus not only led at times to enormous failure—pursuing an invention well beyond the point where it was clear that there was no market for it—but also at other times to enormous successes. His story vividly illustrates what can happen when one's Systemizing Mechanism is ramped up to the maximum and one's Empathy Circuit is tuned low.

Not all scientists and technologists fall at this extreme, yet a strong bias toward systemizing over empathy can be observed in modern-day inventors like Bill Gates, based on his own description of himself in his twenties, when he was founding Microsoft:

"I was a zealot. I didn't believe in weekends. I didn't believe in vacations. I knew everybody's license plate, so I knew when they were coming and going."[20]

Gates's mind had a mental look-up table for detecting *if-and-then* patterns that showed hundreds of employees and their unique car number plate. In the documentary about his life, *Inside Bill's Brain: Decoding Bill Gates*, it is clear that he struggled to understand his mother's feelings and thoughts, even though she tried hard to help him, and that he was an awkward, socially isolated child and teenager. His mother patiently taught him social skills long after the age when his peer group had intuitively mastered these, giving him a set of rules about how to behave in social situations. Although Gates's affective empathy is clearly intact—he gives millions of dollars to alleviate suffering in the poorest parts of the world—these biographical accounts suggest that his social development was delayed, while his systemizing was way ahead of his peers. Steven Levy, reviewing the documentary in *Wired* magazine and having interviewed Gates dozens of times, commented that "Bill Gates arrived on Earth as a Martian." His profile fits that of a hyper-systemizer.

▌▌▌▌▌

Edison famously went through 10,000 loops of checking and rechecking his *if-and-then* patterns to detect important mistakes or to find novel valuable patterns. This is reminiscent of how modern engineers check and recheck manufacturing processes to eradicate one-flaw-in-a-million cycles of the system. Indeed, engineers today repeat the cycle of *if-and-then* patterns not 10,000 times but *one million* times, to ensure that the new system they have created or assembled delivers a near-identical result every time. They call this "*Six sigma*," written like this, using the Greek character sigma:[21]

$$6\sigma$$

It's called *Six sigma* because it is six standard deviations from the mean—an extreme outlier. Engineers, who are hyper-systemizers, want 99.99966 percent of the repetitions of a mechanical system to be defect-free, allowing for just 3.4 defects per million opportunities (DPMO).[22] That's pretty close to that definition of perfection. I find it hugely reassuring that, when the plane I'm on takes off, or when I sit on the chairlift at a ski resort, these machines will work flawlessly at least 999,996.6 times out of a million. The *Six sigma* rule is not just reassuring to us as passengers and consumers, but it can also lead to massive profits. General Electric, for example, announced that when they first used *Six sigma*, their profits grew by more than $1 billion.

Good engineering and invention hinge not just on going through the *if-and-then* steps but also on the feedback loop, which itself entails the twin processes of *iteration* and *refinement.* These map onto steps 3 and 4 in the Systemizing Mechanism (see figure 2.1). Iteration simply means repeat, virtually indefinitely. Refinement is tweaking the system by changing either the *if* or the *and* variables, to fine-tune, optimize, or obtain a new output. The engineer looks at each component in the system and analyzes it for potential weaknesses or—in the worst-case scenario—actual system failure.

It is said that engineers are "omnipresent but invisible": the products of engineering are literally everywhere in human society (and conspicuously absent by and large in non-human animals' worlds, save for limited examples like termite mounds and birds' nests, but where there is no evidence of the individual animal being motivated to experiment with variation and are likely the result of a rigid genetic program).[23] We don't tend to notice engineering until it goes wrong. For example, an estimated 100,000 planes take off and land somewhere in the world every day, but the only ones we hear about are the ones that crash. Thankfully, in 2018 there were only fifteen plane crashes across the world, which equates to one per three million flights.[24] The products of modern engineering are successful simply because they work, and the hyper-systemizing engineers who designed and installed them remain anonymous and invisible.

Many of us have experienced the frustration of a jammed pepper grinder. No matter how hard you try to turn the crank, nothing comes out, as if the wheel of the grinder has stopped working. But often the grinder isn't the problem: the issue is congestion. Vint Cerf, who back in 1973 invented the TCP/IP protocol, a system for electronic communication, became interested in how congestion arose in his pepper grinder.[25] First, he dropped a *handful* of peppercorns into it all at once and saw that the mill got blocked. Then he poured peppercorns in *one at a time,* and they didn't get stuck, but instead flowed out smoothly.

For Cerf, solving the pepper grinder problem was about solving *any* congestion problem where there is variation in the flow across time, whether it is car congestion in cities, the post office not being able to cope with the volume of letters, or email congestion in your online service provider. Cerf's systemizing is how most successful scientists, engineers, and inventors approach problem-solving.

As I think about the complex tools that modern humans make to solve a problem, from the mundane (grinding your

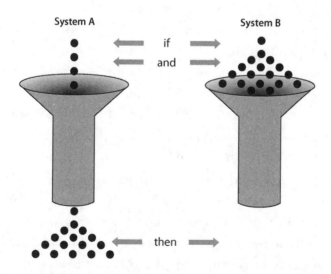

Figure 4.1. Systemizing the pepper grinder to solve the congestion problem

coffee beans in the morning) to the extraordinary (landing a rocket on the moon), the thinking process is exactly the same: *if-and-then* pattern-seeking, followed by the repeating feedback loop.[26] This is the working of the humble, infinitely powerful Systemizing Mechanism that has been inside the human brain for 70,000 to 100,000 years and will continue to deliver invention for our species for a lot longer than that.

||||||

Science and technology are not the only systems-based fields in which we would expect to see the benefits of having a Systemizing Mechanism tuned to a high level. We usually think of STEM fields as ones that might require a systematic mind, but many arts-related fields also benefit from systemizing, such that some now refer to STEAM (the A standing for the arts).[27] Arts such as music, dance, craft, and design are all domains in which we can see the Systemizing Mechanism at work, leading to invention. And as we discussed earlier, even the arts of filmmaking; writing drama, literature, or comedy; or the performing arts can all be systemized, to result in invention.

Consider Glenn Gould, the virtuoso classical pianist, who had an amazing memory for music and was also known for his obsessive practice routine.[28] Going over and over one particular musical phrase or sequence is the *if-and-then* algorithm writ large. He also practiced mentally so that he didn't need access to an instrument. And the inventiveness we see in jazz and other musical composition and improvisation is how the *if-and-then* pattern is modified in step 4 of the process.

As a child, Gould learned to read music before he could read words. His father describes how Glenn wouldn't come out of his bedroom until he had memorized a whole piece of music. As an adult, he had to control every aspect of his life, complaining, for example, about even slight changes in temperature. He played the piano only if he could use his own special chair designed so he could sit very low at the keyboard: he had to sit exactly fourteen inches above the floor. He also rocked back and forth while playing the piano, including during performances. Rocking back and forth is a very physical repeating behavior, a soothing *if-and-then* pattern on repeat. To achieve total control, he gave up live musical performance in favor of recording. He also hated the cold and often wore gloves even in warm places. He hated being touched, refusing to shake hands with people, and hated social functions, and in later life he limited his social contact to letters. In his hometown of Toronto, he would go to the same diner between 2:00 and 3:00 a.m., sit at the same table, and order the same meal of scrambled eggs. Some have speculated as to whether Gould was autistic but was never formally diagnosed. Such speculations may be misguided if he functioned just fine, because a diagnosis is only ever for those who are struggling to cope and are seeking support.

In contrast, Jonathan Chase, a master of the bass guitar, does have a formal autism diagnosis. His approach to music again displays the workings of a hyper-systemizing inventor's mind. Chase talks explicitly about visualizing patterns on the frets of his guitar. He sees the key of C major as a series of dots in the grid of the frets. In Chase's mind, these dots across the frets are joined by imaginary lines, forming an identifiable shape: two sharp spikes. He uses these, and other shapes, to build repeatable *if-and-then* patterns, which he fits together into riffs that he can execute with precision and speed, identically and with perfection, every time. He can go around the loop 10,000 times.[29]

And Chase can vary the series of patterns systematically, so that the result is a seemingly endless ability to improvise jazz. He is producing beautiful, repeatable patterns. *If* he hits the eighth fret on the bottom string, he plays an A, *and* if he moves across to the eighth fret on the next string, *then* he plays a D. Each new note creates a pattern with the note that precedes it, and the sequence of notes in a riff is another pattern.[30]

An obvious place where we can see hyper-systemizing is in the world of games. Max Park is autistic, diagnosed at age two with a delay in social and fine motor development. At age ten, he was given his first Rubik's Cube, and by age fifteen he had won the World Championship in both the 3x3 Rubik's Cube and the one-handed events. His average solve time was 6.85 seconds with two hands, and 10.31 seconds with one hand. He had systemized the 3x3 cube. At best, solving the cube takes a minimum of twenty-two moves. You can see how rapidly *if-and-then* reasoning would help solve the cube: "*if* the red cube with the green side is positioned on the top layer on the right side, *and* I rotate the top layer anti-clockwise by ninety degrees, *then* this will complete the top layer as all one color." Rapid is of course an understatement.[31]

We can also see hyper-systemizing and invention in elite athletes. An example is Los Angeles Lakers basketball All-Star Kobe Bryant, who tragically died in 2020 in a helicopter accident. Bryant looked for patterns in his performance and followed strict regimes. In high school, he practiced basketball moves for fourteen hours a day, from 5:00 a.m. to 7:00 p.m. As a professional athlete, he had a room in his home in which he could rehearse the actions associated with imaginary shots, over and over again, with no distractions—without even a ball or net being physically present. He even figured out that *if* he closely examined the sole of his basketball boot, *and* shaved off a few millimeters, *then* he would achieve an improvement of one-hundredth of a second in his reaction time. Bryant also systemized his hobby of music, learning to play Beethoven's "Moonlight Sonata" by putting a recording on loop and figuring out the composition by ear. Bryant's approach to both basketball and music reveals that his behavior was the product of the Systemizing Mechanism in hyper-mode.[32]

Some hyper-systemizers, in a range of fields, have been described as autistic. For example, Andy Warhol in the field of art, Ludwig Wittgenstein in the field of philosophy, Hans Christian Andersen in the field of literature, and Albert Einstein and Henry Cavendish in the field of physics, have all been described as autistic.[33] In my view, it is unhelpful to speculate if someone—living or not—might be autistic, since a diagnosis is only useful if the person is seeking help and is struggling to function. Diagnosing someone—living or not—on the basis of fragmentary biographical information is unreliable and arguably unethical, since diagnosis should always include the consent of the person and be initiated by them.

And from a scientific perspective, hyper-systemizing does not automatically mean you're autistic. These two descriptions are not synonyms but merely overlap, both in terms of "cognition"—how you process information—and in terms of genetics and prenatal sex steroid hormones (just some of the causal factors). Equally, hyper-systemizing does not automatically make you an inventor or an exceptional musician or athlete. But being a hyper-systemizer increases the probability that you will invent something, because if you keep experimenting with new *if-and-then* patterns, you are more likely to find a pattern that produces a potentially groundbreaking result. Indeed, hyper-systemizers can excel in any field in which they can search for *if-and-then* patterns. Of course, whether your novel system becomes commercially successful depends on whether you also have the opportunity, resources, and skills to exploit your idea. This echoes our discussion way back about the difference between an invention and an innovation, which often needs such resources to be disseminated or become a product that can be taken to market.

We've focused on the Systemizing Mechanism in modern times, but throughout the book I've claimed that the Systemizing Mechanism has a history stretching back 70,000 to

100,000 years and is the result of human evolution. To prove that claim, we need to show that systemizing was absent in our hominid ancestors. It's time to look at three of our ancient human ancestors, *Homo habilis*, *Homo erectus*, and *Homo neanderthalensis*, to establish that there was indeed a revolution in the human brain.

Chapter 5

A Revolution in the Brain

The first stone tools appeared 3.3 million years ago, but for much of the time since then, while there were many species of *Homo*, and their tools showed some changes in complexity, there was in my view no real evidence of a capacity for *generative invention.*

Let's just pick out three among the ancient species of *Homo*, who spanned the last two million years: *Homo habilis, Homo erectus,* and *Homo neanderthalensis.* They could all make simple stone hand hammers and axes to smash, cut, and scrape. But did they invent? When we look at their tools, it gives us pause, because these were *simple* tools. Simple because all you have to do to make one is take a piece of rock and chip at it with another rock. And although, as we'll see, there were *small* changes in these simple stone tools across these three ancestors, for this long two-million-year period there was no really *big* change in the complexity of these tools. Their creators showed no sign of possessing a Systemizing Mechanism.

I'm going to argue that none of these hominids could invent, if we define inventing as being able to come up with a novel tool *more than once.* I use this strict definition (I call it generative invention) because a new tool that an animal comes up with could have arisen as a result of chance (e.g., smashing a nut with a rock), plus "associative learning," the animal then repeating the action sequence because it leads to a reward (e.g., getting the juicy insides of the nut).[1] Associative learning requires a level of intelligence and is widespread in the animal kingdom, but I argue that it's not the same as generative invention.

Let's look at these three ancestors in more detail.

Homo habilis lived in sub-Saharan Africa between 2.1 million years ago and 1.5 million years ago. They produced Oldowan tools, so named because they were first discovered in the Olduvai Gorge in Tanzania. *Homo habilis* were shorter than modern humans, with a brain less than half the size of ours. (Our cranial capacity is 1,496.5 cm^3, whereas theirs was just 610.3 cm^3). But by and large, they kept on making the same tool, over and over again, showing no evidence of a capacity for invention.[2] And these were just very simple tools with not much more than three functions: smash, cut, and scrape.

By contrast, *Homo erectus*, who lived between 2.1 million years ago and 250,000 years ago, had a bigger brain than *Homo habilis* (with a cranial capacity of 1,092.9 cm^3) and was impressive for several reasons. For one thing, these hominids were the first of our ancestors to leave Africa, spreading into Europe and Asia. And they are called *erectus*, for "upright man," because they pretty much gave up an arboreal (tree-dwelling) life, became almost completely terrestrial, and, most importantly, were bipedal. Historian Yuval Harari argues that their hands became more innervated as they developed greater fine motor control. Being upright allowed them to use their hands for other purposes, so they could not only make tools but carry their stone tools with them. And they made a new stone hand ax, known as an Acheulian tool (so named because it was first found in Saint-Acheul, north of where Paris is today).[3] But was this evidence that *Homo erectus* could invent?

Some people argue that they could because, whereas *Homo habilis*'s tools were made by hitting a stone with a second stone as a hammer, *Homo erectus*'s stone tools were made by hitting a stone with different "hammers": a bone, an antler, or a piece of wood. These new hammers allowed for greater precision when making a stone ax. But I would challenge the view that this constituted an invention because, in my definition of the capacity for generative invention, there has to be evidence that an animal can come up with something new repeatedly, *more than just once. Homo erectus* niche was using tools to get marrow out of a bone, just as a woodpecker gets sap from a tree. One might argue that their new tools were inventions because they were using these tools for a new function, but a more parsimonious interpretation is that the same clear reward drove their tool-making behavior, namely, getting a food source out of an object with hard casing. Lots of animals can do that, and creating a one-off novelty is insufficient to count as generative invention, as it can arise by chance plus associative learning.

Finally, take *Homo neanderthalensis* (just called Neanderthals by most people). They lived between 300,000 years ago and 40,000 years ago.[4] They got their name from the Neanderthal region of Germany, where they were first found. Their cranial capacity was 1,500 cm^3, slightly larger than our 1,496.5 cm^3, and they had large brow ridges and a slightly protruding face and chin. Neanderthals used Mousterian stone tools (so named because they were found in Le Moustier, in the Dordogne in France). Their tools were sharper and finer than those of their predecessors, but rather than being an indication of a generative capacity to invent, this may simply reflect their greater strength in being able to grip them. Some argue that they used more complex tools because they may have used fire and made hearths, both good points that we will return to. Neanderthals' tools have been found in the Greek islands, prompting some people to speculate that they were able to invent boats, but a more cautious explanation is that they simply swam to the Greek islands. Some speculate that Neanderthals could produce birch tar as an adhesive, or produced intentional burials, but this evidence is challenged.[5] So again, there is no clear evidence that they could invent.

In sum, my view is that our hominid ancestors were not inventing, if we employ a strict definition that an animal's capacity for invention should be *generative*, stemming from a single drive to experiment, with anything. With a generative capacity, an animal does not just make the same simple stone hammer

Figure 5.1. The first stone tools.

Top: *Homo habilis*'s Oldowan stone hand axes.

Middle: *Homo erectus*'s Acheulian tools.

Bottom: *Homo neanderthalensis*'s Mousterian stone tools.

or ax but is able to come up with hundreds of new designs. In our hominid ancestors, we just don't see this generativity. Genuine invention should be like genuine language: once you can produce one sentence, you can produce hundreds of novel sentences. We would be reluctant to say that a parrot has genuine language if it can only echo the *same* phrase over and over again. In the same way, we should not assume that simply using the *same* tool over and over again, with no new design features, stems from a genuine capacity for invention.

I don't want to belittle what our hominid ancestors were doing, because using a simple stone tool to smash, cut, or scrape is still evidence of the capacity to learn. Our hominid ancestors had learned that using the tool brought rewards, and this alone can explain why they kept making tools. But the capacity to learn is not the same as the capacity for generative experimentation or invention. And many species of animal are capable of learning even if they don't invent.

But then it all changed.

▌▌▌▌

Roughly 200,000 years ago, *Homo sapiens* evolved in East Africa.[6] Archaeologists such as Christopher Henshilwood argue that between 70,000 to 100,000 years ago humans went through a revolution in their tool-making and in our way of thinking: they started to experiment and invent, and to do so generatively. So what do we see in the archaeological record that gives a clue that a big change had occurred?

First, in South Africa, there is evidence of *engraving* from 77,000 years ago, a clear sign that humans were producing specialized tools. Engraving a rock was not an isolated case: soon after, we see engraved ostrich eggshells that are 60,000 years old.

A second clue that generative invention had begun was the discovery of a set of what some archaeologists interpret as beads—the first necklace or *jewelry*.[7] Dating back to

Figure 5.2. Examples of the earliest engraving. *Left*: From 77,000 years ago.
Right: Engraved ostrich eggshell from 60,000 years ago.

about 75,000 years old, one set of shell beads was found in
the Blombos Cave on the southern tip of Africa, overlooking
the Indian Ocean. Again, this was not an isolated case because
another set of such perforated shells dating back even earlier, to 82,000 years ago, was found in Grotte des Pigeons
in Morocco, North Africa. Not all archaeologists agree that
these are beads, but this interpretation is not unreasonable.
The beads were made of snail shells that had been collected
miles away and appear to have then been carefully drilled
with holes. Let's assume this really was a necklace.

And here's a third clue: *bow-and-arrow hunting* is exclusive
to *Homo sapiens* and dates back to 71,000 years ago in South
Africa.[8] We know this because of the archaeological evidence

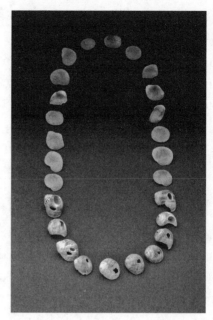

Figure 5.3. The first jewelry, from 75,000 years ago, including ten shell beads. (Facsimiles complete the necklace.)

of mortal injuries inflicted by small obsidian blades that rained down on a group of hunter-gatherers near Turkana, in Kenya, at that time.[9] (Obsidian is a volcanic glass so tough and sharp that, even today, modern surgeons sometimes choose scalpels made out of it.) Bone and stone arrowheads have been found in South Africa dating this far back. The bow-and-arrow likely was used both for hunting and as a deadly new weapon because of its three remarkable advantages: silence, distance, and projectile force, each vital in both hunting and killing. The inventor would have needed to experiment with the best kind of wood for the bow and for the arrow shaft, the length of the cord, and the best material for the arrowhead, to optimize distance and velocity. Again, when you analyze it, constructing a bow-and-arrow requires systemizing: "*if* I attach an arrow to a stretchy fiber, *and* release the tension in the fiber, *then* the arrow will fly."

So, from at least 70,000 years ago—and I've cautiously opted to push the time-window as far back as 100,000 years, as some of the examples above date back to 82,000 years and it's very likely that more such "complex" tools or artifacts will come to light that precede these perforated shells and engravings—we can conclude that humans were no longer just making simple stone hammers, picks, and axes, as our ancestors had been doing for millions of years up until then.

What caused this revolution in tool-making? Why did humans start to invent generatively? My thesis is that the remarkable revolution that shifted humans from simple to complex tool-making 70,000 to 100,000 years ago is best explained by the idea that the Systemizing Mechanism had evolved.

Here's why. Engraving, to me, is a clear sign of *if-and-then* thinking, the defining property of the Systemizing Mechanism.[10] Critically, humans now showed the capacity to think: "*if* I take a smooth stone, *and* use a tool with a fine blade, *then* I can engrave patterns on the stone." In a similar way, to make a necklace also required a mind capable of *if-and-then* thinking: "*if* I have a number of shells, *and* drill a hole into each shell, *and* thread a length of fiber through each hole, *then* I can make a necklace." This simple chain of beads again reveals the capacity for *if-and-then* reasoning.

Recall my strict definition of a capacity for genuine invention: that it should depend on evidence not of a *single* new artifact but rather of the workings of a generative capacity, which is a defining feature of the Systemizing Mechanism. Genuine invention can be said to occur only if we see a blossoming of *many* new artifacts.

In case we needed more evidence of an explosion of generative invention, around the same time, *Homo sapiens* crossed to the Andaman Islands in the Indian Ocean 65,000 years ago, and to Australia 62,000 years ago. This is pretty good evidence that they had invented *boats*, since to reach Australia from the Indonesian islands involved crossing sea channels more than 100 kilometers wide.[11] We also see evidence that 42,000 years ago *Homo sapiens* were able to catch and eat fish.[12] The oldest *fishing hook* is dated 23,000 years ago, while evidence of deep-sea fishing, via analysis of pelagic (tuna) fish bones, is dated to 42,000 years ago. We can only imagine how both a person's hunting success and their health would have been transformed by this invention.

So, after more than two million years of incredibly slow change in tool use, the human drive to experiment and invent suddenly started to flower in extraordinary ways. And the generative property of these human inventions is undeniable: we see deliberately *adorned graves* from 42,000 years ago, and *hand-printing* on cave walls in Indonesia created 40,000 years ago.[13] These are all signs of the same unstoppable drive to invent and experiment generatively that we see in modern humans today but that was absent in all other hominids, or indeed any other animal, before 70,000 to 100,000 years ago. And we see *constructed dwellings*, which were not seen among our hominid ancestors: circular huts in eastern France first appeared around 30,000 years ago.[14]

And indeed, by 23,000 years ago, we see the emergence of another specialized tool: bone *needles,* presumably used to stitch animal hide to make clothes.[15] It is thought that we are the only

Figure 5.4.
Cave art
(hand-printing)
from 40,000
years ago

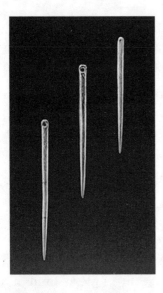

Figure 5.5. Early needles for making clothes, 23,000 to 30,000 years old, from China

species to have ever invented a tool for making clothes. Awls (pointed tools for stitching clothing but which lack "eyes") may date back to 61,000 years ago.

Are we the only species to have shown genuine invention? Some argue that a hand-print in a cave in Spain dating from 64,000 years ago and an etching in Gibraltar dating from 34,000 years ago could be Neanderthal in origin.[16] However, interpretation of these remains controversial. A conservative view is that only in *Homo sapiens* do we see strong evidence of genuine invention, from *many* novel artifacts, all of which can be unambiguously attributed to us.

Each of these new inventions by *Homo sapiens* was, in my argument, the expression of a mind playing with *if-and-then* patterns—the workings of the new Systemizing Mechanism. Consider how cave painting entails various types of *if-and-then* reasoning: the invention of a way to make marks on the wall (*"if* I have some yellow ochre, *and* make a mark on the wall with it, *then* a yellow mark stays on the wall"), followed by experimenting with making marks on walls in a systematic way (*"if* I make marks on the wall, *and* do them in this particular order, *then* the marks have the shape of a bison"). And we see the same experimenting with patterns in Germany 35,000 years ago, when humans were carving sculptures: an extraordinary ivory figure of a "lion-man" and a "Venus figure" of a woman with exaggerated sexual features.[17]

At the most basic level, inventing a sculpture like the lion-man involved reasoning along the lines of: "*if* I take the shape of the top half of a lion, *and* combine it with the shape of the bottom half of a human, *then* I can make a lion-man (sculpture, drawing, word, image in my mind)." Of course, there's a lot more than systemizing involved here, and we'll come back to

Figure 5.6. The earliest sculptures.

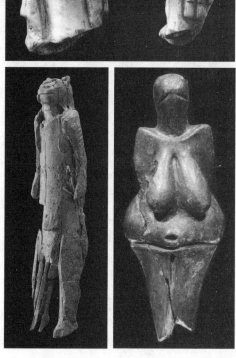

Top: Mammoth ivory Venus figurines, from 25,000 years ago.

Left: Ivory sculpture of a lion-man from 32,000 years ago.

Right: Ceramic sculpture of a Venus figure from 29,000 years ago.

this in chapter 7, particularly the idea of being able to imagine, and to imagine fictional entities. Alongside having the idea of making a sculpture or painting, one has to be able to make the tool that will be needed to carve the sculpture or paint the painting. Complex, specialized tools. I argue that the Systemizing Mechanism was a prerequisite to be able to invent.[18]

Indeed, to me, the most compelling evidence that a cognitive revolution occurred in the human brain is the sudden change in the rate of invention in the time line of tool production, from almost flatlining for 2.6 million years to reaching a tipping point between 70,000 and 100,000 years ago.[19]

Yuval Harari also pinpoints the date of the cognitive revolution to 70,000 years ago. I am adopting a slightly more generous view that the cognitive revolution dates as far back as between 70,000 and 100,000 years ago, because my reading of the archaeological evidence is that the cognitive revolution occurred gradually over this period, and we can anticipate that new archaeological discoveries of complex tools will emerge in this time window. Archaeologist Richard Klein argues for a genetic mutation 40,000 to 50,000 years ago that caused an

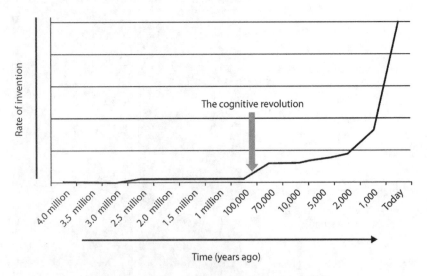

Figure 5.7. The cognitive revolution in tool-making

abrupt change in human cognition and behavior.[20] I would agree with this choice of the date because, from then on, the evidence for invention is much clearer. But it is unlikely that a single genetic change can explain the evolution of the Systemizing Mechanism. Some big genetic changes may have contributed to the evolution of the Systemizing Mechanism, but much more likely is that hundreds if not thousands of common genetic variants contributed to its evolution. We'll look at the genetic evidence later, but suffice it to say that polygenic traits (traits involving many genes, each having a small effect) typically evolve gradually, not abruptly.[21]

Taking the wider time window of around 100,000 years ago for the cognitive revolution, it is interesting to speculate on the influence of the cognitive revolution on the two major dispersals by humans out of Africa—one 108,000 years ago, when they went to the Levant and coexisted with the Neanderthals, and the other 50,000 years ago that led to humans replacing the Neanderthals by 40,000 years ago.[22] Humans ended up in what today we call Australia by 40,000 years ago and in what today we call North America by 16,000 years ago. Were these dispersals out of Africa and across the world caused by the cognitive revolution as humans began to be able to invent new complex tools (like boats) and complex ways of understanding the natural world (like reading the stars) to navigate continents?

My claim is that these more complex and specialized tools were emanating from the Systemizing Mechanism. This new pattern-seeking engine in the mind used a generative algorithm that could come up with and test an infinite number of new *if-and-then* patterns, unleashing invention at a runaway rate. It enabled humans to reason "*if* I take x, *and* make one change to it, *then* x becomes y." Humans had become pattern seekers, of a very special kind. We went from being able to make a simple stone hammer or ax to being able to invent anything. And the Systemizing Mechanism also allowed the first humans to do

Figure 5.8. The 40,000-year-old bone flute

something else rather remarkable. We—and no other animal before or since—invented rhythm and music.

About 40,000 years ago, someone in Germany picked up a bone and turned it into a flute—to date, the oldest musical instrument in the history of the world. The Systemizing Mechanism was responsible for making the bone flute (a tool for making music), for making the tool to make the bone flute, and for making music itself (a tool for experimenting with sounds).[23] All of these are complex tools, each a system characterized by *if-and-then* patterns. Music is, at one level, nothing more than a sequence of (rhythmic and tonal) patterns that we can intentionally vary, using *if-and-then* rules, although as we all know, its effects can be extraordinarily emotional. But prior to being able to have the emotional experience, we first need to be able to recognize music as patterns.

To me the bone flute is a beautifully clear illustration of the Systemizing Mechanism in action. The person who, 40,000 years ago, picked up this hollow bone with a hole in the side may have asked themselves a question: "What sounds can I make with this?" As they blew into one end and then placed a fingertip over the hole to cover it, they heard how the sound changed. One action (covering the hole) seemed to make a new sound, and another action (uncovering the hole) seemed to make the original sound. So this person hypothesized an *if-and-then* pattern: "*if* I make a hole in the bone, *and* I cover the hole with my finger and blow, *then* I make a different sound." And they tested to see if this *if-and-then* pattern held true by repeating it over and over again.

Then the maker of the first bone flute made a second hole in the bone and repeated the whole systemizing process, listening to what happened when they blew and covered one or both holes separately or together. And then they must have done this on another hollow bone, and then on another, to experiment with another *if-and-then* pattern, varying the distance between the holes, until they found the combination that sounded most appealing.

When I read about the bone flute, I immediately emailed archaeologist Nicholas Conard, director of the beautiful little museum in which it is kept in Blaubeuren in Germany, to ask him if I could come visit to have a closer look at it. To my excitement, he emailed me back almost immediately, suggesting we first meet at the Hohle Fels cave, so I could see exactly where the bone flute had been found and where our ancestors had lived.

I flew to Stuttgart and took a taxi deep into the countryside to Hohle Fels. As I went inside the dark cave, it felt like entering a time machine, going back to witness the origins of art and music. From the cave mouth, I descended a ladder to the deepest level of the cave. There was Nicholas, who shook my

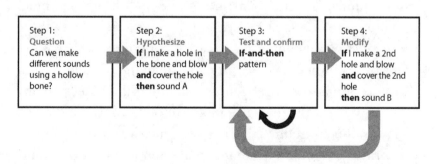

Figure 5.9. Inventing a musical instrument using the Systemizing Mechanism

hand and welcomed me warmly. He pointed up to one stratum of the rocks and said:

"Right here is twenty thousand years ago."

Then he pointed about a meter lower, to the next stratum in the rocks, at ground level, and said:

"Here, we are at forty thousand years ago."

I felt a shiver of excitement. I looked down at my feet, in my modern leather shoes, suddenly aware that I was standing exactly where early humans had stood, sat, slept, and eaten 40,000 years ago. I tuned back into the present: Nicholas was showing me how he and his team were painstakingly combing through thousands of fine stones, in search of what might be tiny fragments of bone or other durable materials that could have been transformed into a human artifact or tool.

Then we drove to the museum. The bone flute is as thin and as small as your baby finger, made from the hollow wing bone of a griffin vulture. I looked at the holes along the bone flute and again had that powerful feeling of connection with the past: its maker had put his or her fingers on these holes. Nicholas played me a recording of the bone flute, performed by a modern musician, and I realized that its maker had a musical ear similar to ours today: he or she had drilled *five* holes into the piece of bone, spaced apart in a pattern, such that the flute would play a pentatonic scale.[24] A pentatonic scale has five notes per octave; developed in many ancient civilizations, it is the basis of many musical genres, including the blues and jazz that many of us still enjoy listening to 1,600 generations later.[25]

From the museum I sent a text message to my son in Cambridge to communicate my excitement, and almost immediately he texted back. I reflected on how the same Systemizing

Mechanism that enabled the invention of the bone flute was also responsible for inventing a text message.

And I reflected on how we are the only species that produces and responds to music, narrowly defined as rhythmic, harmonic, and melodic sound patterns that are intentionally varied in a systematic way.[26] Music is able to have a deep emotional effect on our brain because we use a Systemizing Mechanism to recognize that someone else is intentionally varying these *if-and-then* patterns (if we're the listener) or to produce them ourselves (if we're the musician), to influence our emotions. Studies involving fMRI scans of the brain confirm that we find listening to music pleasurable, in that the ventral striatum, part of the reward circuit of the brain, is activated when we listen to it. We need the Empathy Circuit to intentionally influence another person's emotional state through music or to imagine the emotional communicative intention of the composer, but we need the Systemizing Mechanism to recognize and produce the *if-and-then* musical patterns.

Many animal species call to each other, and some of these calls can be described as song because they have melody. Perhaps the best studied is birdsong. There's no question that birdsong has an emotional impact on the bird-listener.[27]

For example, if you put a female sparrow into an audio-booth and play a recording of a male sparrow's mating call to her, there are elevated expression levels of a gene called Egr-1 in the reward pathways of her brain. Egr-1 expression levels in these reward pathways are also elevated when an animal is given a different kind of reward: cocaine. Remarkably, if just before hearing the song the female sparrow is given a little capsule containing either estradiol (an estrogen) or placebo, to mimic her hormone levels in the mating season, she expresses even more of the genes in the reward pathways of her brain when she hears the male sparrow's song. So, for a female sparrow, we can infer that hearing the male mating song is pleasurable.

In contrast, if you put a *male* sparrow in the audio-booth and play the same recording of a male sparrow's mating call to him, the same gene expression profile is not seen in the reward pathways of his brain. Instead, he shows it in the amygdala, a brain region associated with detection of threat. If the male sparrow is given a testosterone capsule just before hearing the male song, so that his hormone levels are equivalent to those in a typical male during the mating season, this brain response to hearing male birdsong is intensified. (And this is not seen to the same degree when he's given a placebo.) For him, we can infer that the male mating song is not pleasurable but rather is a warning signal that another male is challenging his territory.

So, in another species, song can affect the listener's brain positively or negatively. But that's a far cry from saying that birds or any other animals identify such sounds as *music*.[28] Recall my definition of music: intentional, systematic variation of notes or the beat to explore *if-and-then* sound patterns. If new, this is generative invention of sound patterns. In contrast, most birds just produce the same melodic sequence, without much variation. Primatologist Valerie Dufour reviewed the evidence for whether, when chimpanzees drum by beating their hands against tree roots or against their bodies, this counts as music.[29] She concluded that it lacks isochrony (or evenness), a key feature of music. And there's no evidence of it being intentional *systematic variation* of the beat, to explore *if-and-then* sound patterns. The conclusion from the study of other animals is that they just don't get the beat in music.

In contrast to what is observed in other animals, young human children are magnetically drawn to music. Give a toddler a drum or a keyboard, and he or she will imitate or generate a musical sequence, and then *vary* the sequence.[30] In the absence of a drum, you can just clap a rhythm, and a toddler will pick up the rule and try to clap it back or vary it. Young children are listening out for *if-and-then* sound patterns, but there is no evidence that apes or monkeys are.[31] One study found that dogs in kennels show reduced stress behaviors in response to music, but this doesn't prove that they get the beat or recognize other musical patterns.[32] And although canines (dogs and wolves) may howl together, it is unclear that they are actually trying to produce music. Other animals don't respond to human music as music, narrowly defined.

But back to our hominid ancestors: there is no solid evidence that Neanderthals or any of our other hominid ancestors made music either.[33] According to archaeologist Steve Mithen, Neanderthals did not produce a musical instrument. Interestingly, this was challenged in 1996 when archaeologist Ivan Turk announced that he had found an instrument in a cave in Slovenia. It was a bear's thigh bone, pierced by two circular holes, and Turk said it was a flute. But subsequent evidence cast doubt on this in several ways. First, another archaeologist, Francesco D'Errico, found more bones in the same cave in which almost identical holes had been chewed by carnivores, and tooth marks opposite the holes suggested that the bones had been gripped by jaws. So the perforated thigh bone was not the result of a Neanderthal *deliberately* making holes to turn the bone into a flute for the intentional variation of sound patterns. Most damning of all, the ends of the bones were still blocked by bone tissue. That meant that they could never have been used as musical instruments—no one had been blowing down the hollow bone. In short, the current evidence from our hominid ancestors suggests that music, and the Systemizing Mechanism that enabled it, is unique to humans.[34]

About 43,000 years ago, another human picked up a baboon bone in the Lebombo Mountains between South Africa and Swaziland and invented something else quite extraordinary: a

Figure 5.10. The Lebombo bone with twenty-nine tally marks on it, from 43,000 years ago. Thought to be the earliest counting device.

tool for *counting*.[35] Over time he or she made twenty-nine marks (or notches) along the bone, and archaeologists speculate that these might have been produced for some kind of ritual or, most likely, were a form of tally: a way of counting. Counting, and mathematics more generally, requires *if-and-then* thinking. Consider, for example: "*if* I take a smooth bone, *and* make a mark on it each morning, *then* the bone can show me how many days since the last full moon." (We can easily imagine other early uses of counting, such as keeping a tally of how many items had been traded.)

The Systemizing Mechanism was not just spotting *if-and-then* patterns in music or numbers. Its power was that it could be applied to an infinite number of objects, events, or sets of information, systemizing them, making the mechanism generative and therefore very useful. Humans could now build on or modify previous tools they had created, enabling them to run new iterations of modifications. Consider: "*if* I take this tool, *and* change this one variable, *then* I can produce a new version of the tool." This alone would have led to runaway invention. This transformative new form of pattern-seeking, a new algorithm in the human mind, set us apart from every other animal, and inventing became unstoppable.

For example, by 12,000 years ago, agriculture was invented and would unfold over time. Agriculture could provide food in large quantities because we could now see plants and animals themselves as complex tools or systems that could do the work to feed humans. The story of agriculture—or the domestication of plants and animals—is extraordinary. It was invented independently in multiple locations: it started in southeastern Turkey, the western Levant, and the Levant, and then appeared in China, Central America, and South America. Growing wheat 9,000 years ago meant for the first time we could make bread, and new foods were produced at scale. Soon we were growing peas, barley, and lentils (8,000 years ago); olive trees (5,000 years ago); grapes, cashew nuts, rice, maize, corn, potatoes, and millet (3,500 years ago). We were also domesticating horses, camels, sheep, and goats by this time.[36]

And it is clear that agriculture needed the Systemizing Mechanism.[37] Consider these examples, where one could use one or more of *ands* in the *if-and-then* pattern: "*if* I sow my wheat seeds, *and* I use a hoe to plant them more deeply, [*and* I weed the field, *and* I guard my fields against parasites, *and* I water my field every day, *and* I scatter dung on them], *then* I will get a better crop." (As you can see, you can keep adding as many *and* variables in the *if-and-then* pattern as you wish.) And agriculture was not just about gaining control over plants but also over animals. Consider: "*if* I take my ox, *and* castrate him, *then* he will be more obedient." Or consider: "*if* I take my male chicken, *and* castrate him, *then* the meat quality will be better." In short, agriculture was a sign that humans had systemized nature, harnessing it under their control.[38]

Systemizing continued to be an unstoppable force for generative invention. Consider just four examples of transformational systems developed by humans between 5,500 and 4,000 years ago:

First, the **wheel**: "*if* I take a piece of wood, *and* cut it in a circular shape, *then* it will rotate."[39] The earliest example of a wheel was found in Mesopotamia in the late Neolithic era. Starting about 11,000 years ago and lasting until about 5,500 years ago, the Neolithic era (or New Stone Age) was remarkable because it marked the transition from a hunter-gatherer lifestyle to agricultural settlement and early civilization. The wheel played a vital part in all of this. Think of the many applications of the wheel, all of which are tools that do work for us: the ship's wheel (to steer a vehicle), the potter's wheel (to create a clay pot), the flywheel (to cast a fishing hook), and the cogwheel (to drive another wheel), among many variations.

The second transformational system was **writing**, invented 5,500 years ago in Sumeria in the Euphrates Valley.[40] Writing on clay tablets was first used for counting (to keep track of who had paid tax or who had an IOU against produce). But by 3,000 years ago, cuneiform or Egyptian hieroglyphics was a "full script," meaning that they could be used to convey anything. Once again, we can see the *if-and-then* algorithm at work: "*if* I start with a clean surface, *and* make marks on the surface, *then* these marks can stand for objects or ideas." Of course, this algorithm presupposes a capacity for reflecting on and representing ideas, for which the Empathy Circuit was needed. But the mechanical act of writing also required the Systemizing Mechanism. Writing is so important that it defines the separation between prehistory and history.

Third was **mathematics**, again invented 5,000 years ago.[41] Mathematics of course covers not just arithmetic but all its related remarkable branches and applications: from algebra, geometry, and astronomy to engineering, taxation, and recording time. Consider: "*if* I take the number 3, *and* cube it, *then* the number becomes 27."

And finally: **religion**.[42] We know that Hinduism, for example, is at least 4,000 years old. It included an elaborate system of rules governing the caste system, in which there were *if-and-then* laws of purity and pollution, and distinctions were made between groups of people: the Brahmins (the priest caste) and the Shudras (the servants) were just a few of the 3,000 castes, defined in contrast to the "outcastes" or Untouchables. While the Empathy Circuit would have been needed to imagine a god with a mind, with thoughts and feelings, the Systemizing Mechanism was needed to create a system of *if-and-then* laws.

By 4,000 years ago, cataloging had been invented in many places (Sumeria, Egypt, ancient China, and the Inca empire in South America) so that written information could be archived and retrieved. By 3,700 years ago, the new Systemizing Mechanism allowed us to forge new materials, like bronze.[43] By 1776 BC, King Hammurabi of the first Babylonian dynasty had written a legal code that included rulings formulated using the *if-and-then* format. For example: "*if* a superior man strikes a woman of superior class, *and* that woman should die, *then* they shall kill his daughter."[44]

So systemizing was leading not just to mechanical inventions but to inventions of any system, including legal systems that prescribed a moral code and defined justice.[45]

▌▌▌▌▌

Let's go back to a fact that challenges the Systemizing Mechanism theory: *Homo erectus* used fire, long before the cognitive revolution, at least 400,000 years ago, and on some accounts on a daily basis by 300,000 years ago. Neanderthals used fire too. Surely, some would argue, that's a sign of the capacity for invention and experimentation? If *Homo erectus* was using fire that long ago, doesn't this disprove the theory that the capacity for invention only evolved 70,000 to 100,000 years ago with the Systemizing Mechanism?[46]

It was undoubtedly a huge step forward when our ancestors first controlled fire. The domestication of fire had so many remarkable advantages. First, it enabled our ancestors to eat foods that would otherwise have been inedible (think of rice or potatoes), so they developed a more varied diet. In addition, cooking made meat more tender, enabling it to be eaten more quickly (in minutes, whereas chewing raw meat can take many hours).[47] And eating cooked food is thought to have led to a reduction in human gut size, freeing up energy for growing a larger brain. Fire also enabled our ancestors to stay healthy by eating foods that were free of infectious bacteria and parasites (since these are destroyed by heat during cooking). Fire also allowed our ancestors to clear forests to attract game to graze, which was an easier way to get lunch. The first barbecues would have been our ancestors' cooking what they caught. Aside from cooking, fire most obviously provided our ancestors with light, enabling them to live in caves and see in the dark, as well as to scare away lions and other predators. And it enabled our ancestors to stay warm in colder environments as they moved out of Africa.

Later, the domestication of fire would lead to a suite of other things, like baking clay to create ceramic objects, heat-treating stones to create stronger tools, and clearing forests for planting. But these last functions of fire would not have been relevant to *Homo erectus*, because there is no evidence that they engaged in any of them. So, although *Homo erectus* used fire, they may not have been systematically experimenting. Rather, they were likely using it just opportunistically, keeping naturally occurring fires burning, for example, after lightning strikes. In other words, using fire need not be evidence that *Homo erectus* invented fire production.

Controlled use of fire is more relevant to the Systemizing Mechanism, since it suggests that fire was used for a purpose, *systematically*, and as a tool. The earliest evidence of the controlled use of fire is from much later, about 200,000 years ago, when we see examples of building a hearth (a fireplace). These were often formed by collecting stones into a ring, perhaps to contain a fire, or perhaps to reuse the same location for a fire. This example of using fire precedes the cognitive revolution by at least 100,000 years. Although it can be argued that building a hearth would have required *if-and-then* reasoning, the fact that this is an isolated example means that it could have arisen, like other forms of animal "culture," through processes like imitation. Recall that for something to count as genuine invention, it should occur *more than once*, in the context of other inventions, not just as a one-off piece of behavior.

In contrast, a clearer example of the controlled use of fire is the creation of earth ovens (hearths that are banked up with clay, sometimes into a dome) for cooking, heating, and, in the first kilns, hardening clay figurines. But these uses are only seen from 40,000 years ago, that is, *after* the cognitive revolution, and they coincide with more intentional uses of fuels (such as wood, peat, animal dung, or straw).

Therefore, while controlled use of fire is an indicator that a big cognitive change had occurred in humans, *simple* use of fire may not be. If we are being super-cautious and not ascribing the capacity to invent or experiment to an animal without *multiple* examples in support of this, then we should conclude that *Homo erectus* simple use of fire could have just been the result of associative learning following a one-off chance event, but not a sign of the capacity for genuine invention.

Was the fact that *Homo sapiens* alone could invent and experiment generatively one of the reasons why Neanderthals became extinct 40,000 years ago? Neanderthals and modern humans coexisted for more than 5,000 years, and from the archaeological record it is clear that we overlapped in time and sometimes even in space, sharing caves. We even had sex together: How else are we going to explain that 1 to 4 percent of non-Africans' DNA is Neanderthal in origin?[48] But looking at the tools Neanderthals used, one can see that these were still mostly the same limited simple tools for smashing, cutting, and scraping while modern humans were producing complex tools like bow-and-arrows and creating art, sculpture, and musical instruments. There is some speculation that Neanderthals may have used spears and possibly adhesives to attach a stone tool to a spear (so-called hafting), but the evidence, which remains incredibly sparse and open to other interpretations, has been challenged.[49] Conservatively, one would have to conclude that there is no good evidence that Neanderthals could experiment and invent generatively.

Neanderthals started to die out around 40,000 years ago, when the cognitive revolution was in full swing.[50] They may have died out because *Homo sapiens*, using our suite of smarter, more complex tools, themselves products of the Systemizing Mechanism, could acquire resources more efficiently, leaving poor Neanderthals with less. And they may have died out because *Homo sapiens*, using our greater social intelligence, including the capacity for deception (a product of the Empathy Circuit in our brain), could run rings around them. There's no evidence that Neanderthals could deceive, and if this was the case, they would have been at a massive competitive disadvantage. As we mentioned earlier, there is no evidence of the stealthy use of arrows or darts by Neanderthals, unlike *Homo sapiens*, who lived at the same time and whose stone tips have endured. Nevertheless, the absence of evidence is not evidence of absence.

Other reasons have been postulated for why Neanderthals died out, but my contention is that the combination of the new Systemizing Mechanism, enabling complex tool-making, and the Empathy Circuit, enabling complex social interaction and deception, led *Homo sapiens* to become unrivaled. The absence of genuine invention in pre-humans and its presence in the human archaeological record point to the evolution of the Systemizing Mechanism.

||||||

Let's go back to the inventor of early jewelry around 75,000 years ago: he or she may have been thinking about how they would be perceived by others (will wearing jewelry enhance my beauty or my social status?), or thinking of someone who might appreciate the necklace as a gift.[51] Either motivation would have required empathy. In the same way, we can assume that a cave painter 40,000 years ago had a theory of mind, if their motivation was to communicate an idea to an imagined audience who might look at the painting at some point in the future. So, the invention of jewelry is a clue that around 75,000 years ago humans were not only capable of *if-and-then* thinking but also that they were *self-conscious*—they were thinking about how others saw them. That little string of beads is a big clue that humans 75,000 years ago were capable of *self-reflection*—another benefit of the Empathy Circuit.[52]

The Systemizing Mechanism gave rise to the capacity for invention and experimentation, while the Empathy Circuit gave rise to the capacity for thinking about another person's thoughts and about one's own thoughts and allowed for flexible communication. Together, these two new cognitive modules constituted the cognitive revolution. Although the Systemizing

Mechanism and the Empathy Circuit are separate, as demonstrated by the fact that many autistic people struggle with theory of mind and yet can be talented at experimenting, these brain mechanisms clearly interact, as is seen in two uniquely human behaviors: language and music.[53]

Systemizing allows us to understand and produce syntax and the other rule-based patterns of language and to recognize and produce the melodic patterns in music. But empathy allows us to read between the lines in language, to understand a speaker's *intended* meaning behind what is or is not said, or is said obliquely or figuratively; and empathy allows us to connect with others emotionally through music. In sum, empathy and theory of mind can explain *why* early humans experimented to make jewelry, art, sculpture, and music, but by itself it cannot explain *how* humans experimented to make jewelry and other forms of art. For that, we needed the Systemizing Mechanism.

Systemizing seems to have been absent in our hominid ancestors. But to really prove that the Systemizing Mechanism was part of the cognitive revolution in the human brain we need to show that systemizing is absent in other animals. To do this, we need to examine the evidence from comparative psychology.

It's time to look at our close relatives—monkeys, apes, and other animals.

Chapter 6

System-Blindness: Why Monkeys Don't Skateboard

If the Systemizing Mechanism in the human brain was the result of a genetic change in human evolution 70,000 years ago, and if the Systemizing Mechanism is what underlies the capacity for invention, leading humans to cross the Rubicon dividing us from all other animals, then we should see an absence of invention in other species alive today.

What about other animals' ability to make tools? Surely that's a sign of their capacity to invent? There's no question that other animals can make and use simple tools, yet it is striking, to me at least, that they don't experiment and invent in the strictest sense of these words—systematically and generatively.[1] Chimpanzees and humans split from our common ancestor eight million years ago, so they've had as long as we have had to develop a capacity to invent complex tools, like a bicycle, a paintbrush, or a bow-and-arrow. But while their lifestyle today is largely unchanged from 70,000 to 100,000 years ago, ours, with all of our inventions, would be unrecognizable to our ancestors.

So, what do we mean when we say that other animals can use simple tools? Monkeys, apes, and even crows and elephants can all use a rock to hammer open the shell of a nut to get the food hidden inside. Each of these animals can also use simple tools in ways that might make you think they have the capacity for invention, such as using a stick to pull some out-of-reach item of food into reach. Chimps can use sticks to forage for ants. Crows have learned to drop nuts onto roads so that cars drive over them, cracking them open to reveal their tasty contents. And they have also learned to drop their nuts on pedestrian crossings, so that when the traffic lights turn red they can eat the crushed nuts without getting run over. Crows have even learned to drop stones into a glass to raise the height of the water, enabling them to grab a piece of meat as it floats upwards.[2]

But to my mind none of these examples of tool use are any more complex than the reports of blue tits pecking open the foil tops of the old-style milk bottles in the United Kingdom that were left on doorsteps, to drink the cream. Evolutionary biologist Kevin Laland described this as an example of animal *innovation,* but for me this is an overly rich account of what is going on.[3] Blue tits could easily have started pecking open milk bottle tops through simple associative learning, plus impressive *social* learning as this new behavior rapidly spread through the blue tit population.[4] Apes do not manufacture complex stone tools in the wild, and attempts to teach them to make such stone tools have failed.

Laland lists a set of other animal behaviors that to him also seem like innovations, such as orangutans using tools to extract palm hearts from trees with sharp spines, or how they make whistles out of leaves to ward off predators, or use branches and leaves to fan themselves, and to scoop to get honey, or how herring gulls kill rabbits by dropping them onto rocks from a great height.[5] But again, these can all be more parsimoniously explained in terms of associative learning.

I've seen a dog open a door by standing on its hind legs and swiping the handle with its paw. Tempting as it is to interpret this or any of the previous examples of tool use as a sign of invention, these actions need not involve *if-and-then* reasoning. Whenever we see an example of such animal behavior, we need to take into account the history of the behavior, which could have first occurred by chance through associatively learning that A is followed by B (for example, hitting the door handle with a paw and the door opening). Then the animal would have repeated the action because it led to a reward (the dog gets out for a run in the garden).

Given that this particular dog didn't show any other examples of apparent invention or experimenting, the cautious interpretation is that his door-opening behavior was just the result of associative learning, not a sign of genuine capacity for invention or experimentation. The pioneer of animal behavior, B. F. Skinner, famously argued that animals can be rewarded to learn sequences of behavior that look quite impressive, like bears riding bicycles in circuses, or pigeons playing Ping-Pong, but learning a sequence of associations still doesn't add up to the capacity to invent.

You can easily see why Laland and others are tempted to interpret some animal tool use as examples of inventions: bottlenose dolphins can carry marine sponges in their beaks to stir up the ocean bottom sand, to uncover where their prey is hiding. Some have speculated that these dolphins are also carrying the sponges to protect their noses from getting scratched while digging on the ocean floor. However we interpret their behavior, at a minimum these are seemingly new examples of tool use. They can also use a conch shell to catch a fish and empty it into their mouth, as though from a cup.[6] Sea otters use stones to hammer shells off rocks and to crack open the hard shells of prey.[7] Gorillas use branches to test the depth of water and use the trunks of shrubs to make bridges across deep swamps.[8] You might also be amused to learn that macaque monkeys in zoos even use visitors' hair to floss their teeth.[9]

Some birds, such as those in the corvid family (crows are an example of a corvid), can even use tools sequentially, using a shorter stick to obtain a longer stick to reach the bait.[10] Elephants impressively cooperate with other elephants to pull a rope together to get food. Elephants also use branches to swat flies and to scratch themselves.[11] Here's my favorite example: octopuses use the empty halves of coconut shells as portable armor, or even as a vehicle they can use to ride across the seabed.[12] And here's my most shocking example: some birds of prey (raptors) even use fire. For example, in Australia black kites (also known as firehawks) pick up smoldering sticks from bush fires, carry them away, and then drop them in dry grassy areas to start a new fire. This causes field mice to run from the fire, and the kites then swoop down from a high branch and catch and eat the mice.[13] All these examples of tool use are undoubtedly very impressive.

Recall the dog who could swipe a door handle (A) and it opened (B). As we discussed, associative learning could lead the dog to pair A and B, because B is rewarding. In contrast, when you or I approach the same problem, we reason: "*if* the door is closed, *and* I turn the handle so the lock slides out of the groove in the door frame, *then* the door will open." When the door fails to open, we systemize the door handle by checking the *and* variable ("Is the handle connected to the lock? Does the lock need some oil to lubricate it so that it can slide back and forth?"). Seeking to find explanations shows up as experimenting, until we have identified the relevant causal variable. Sometimes we call this "troubleshooting," and this may assume some knowledge of the system. But even the search for relevant variables (in the absence of knowledge) is a sign of trying to systemize. We just don't see this kind of behavior in other animals.

Figure 6.1. *Top row*: Many animals can use simple tools. *Bottom rows*: Apparent animal invention.

Top left: A chimpanzee using a rock to crack a nut. *Top right*: A crow drops a stone into water to make the water level rise to get the food. *Bottom left*: Bottlenose dolphins use a conch shell to catch fish. *Bottom right*: An octopus using an empty coconut shell as a vehicle. *Last row*: Firehawks.

The absence of systemizing in other animals is striking. We don't see monkeys and apes adding spice or other ingredients to their food, to experiment with taste. Nor do we see them working out new moves on a bouncy trampoline-like surface or building seesaws, to experiment with movement and causality. It is impressive to watch arboreal (tree-dwelling) monkeys swinging from branch to branch, but they could just be learning through association which size branch to grasp and which to avoid. If they understood the *causal* concepts of weight and support, we would see them experimenting with homemade seesaws or other kinds of balance problems. Although one study showed that chimps are more likely to choose the lower cup on a seesaw as the one most likely to contain food, this may simply reveal that they can learn a rule and draw an inference; we still don't see them experimenting with the causal property of weight in the wild.[14] Nor do they work out moves on a skateboard or even experiment with one as a form of transport.[15] We don't even see them throwing things to see if they travel straight, curve back like a boomerang, or glide like a frisbee, experimenting with object motion. Primatologist Marc Hauser and his colleagues have argued that chimps don't throw things because the morphology of their hands isn't suited to accurate throwing, but humans throw things no matter how deformed their hand may be—humans experiment.

Nor do monkeys and apes dance or experiment with rhythm, yet every human is moved to tap along to a favorite song or get up and dance if they're in the mood. And we don't see apes and monkeys surfing the waves, experimenting with another form of transport. This suggests that, unlike us, they have no curiosity that impels them to experiment, to seek out, to play with *if-and-then* patterns. They see the same information we do—the wave changing shape, the seesaw going up and down—but just ignore it, because they lack the Systemizing

Mechanism in their brain. They are *system-blind*—they don't do any systems-thinking.[16]

▌▌▌▌

If monkeys and apes are system-blind, we should expect them to fail causal *if-and-then* systemizing tests that even a two-year-old human can pass. In fact, even a nine-month-old human infant can recognize causal relationships, but can monkeys and apes?[17] Are they stuck in their own version of the Stone Age that kept our hominid ancestors using only simple stone tools before the cognitive revolution 70,000 to 100,000 years ago? Surely an ape understands causality when it "hammers" the shell of a nut to get the fruit inside? And isn't there some implicit understanding of causality in a chimp's mind when it makes nests out of branches to sleep in at night?[18]

Primatologist Daniel Povinelli investigated this by giving chimps simple causal systemizing tests in the form of building blocks to stack and balance, some of which had extra weights hidden in them. He was effectively asking our primate cousins to figure out how to balance a little model seesaw—whether to put the fulcrum in the center of the plank or more toward one end, if the plank (unknown to them) was unequally weighted toward one end. While a human child can figure out that, in order to make their construction balance, they have to move the blocks with the hidden weights to accommodate their unusual center of gravity, the chimps failed and gave up. They just weren't interested in playing with the blocks to figure out the system behind them. They didn't seem *wired to experiment* out of curiosity, unlike what we humans do all the time.[19]

Povinelli then set the chimps three other tests to really nail down whether they were capable of causal *if-and-then* reasoning. In one test, when chimpanzees were offered the choice of using a flimsy rubber rake or a stiff one to retrieve an item of food, they showed no preference. This suggests that they didn't really understand the causal implications of the more effective stiff tool. In a second test, chimpanzees were offered a choice of pulling a rake that would cause food to fall into a trap and one that would not. Again they showed no preference, suggesting that they didn't understand cause-and-effect. And in a final test, they were offered an inverted rake or one that was the right way up with which to retrieve food. Again, they showed no preference, even though the latter would have been far more effective.[20] It is interesting that chimps reared in human culture can learn some causal reasoning. For example, they have been successfully taught to chip stone tools.[21] But in the wild chimps show few if any signs of understanding causality, suggesting that this is not part of their natural repertoire.

This led Povinelli to conclude, "Chimpanzees consistently focus solely on the observable relations and fail to cognize the unobservable causal mechanisms at stake."

Other primatologists, such as Josep Call, have challenged Povinelli's conclusion, arguing that great apes can draw some inferences under lab conditions and some of these inferences may be causal ones, although the evidence remains open to other interpretations.[22] But if they can, the question is, why don't we see them using this capacity in the wild?

Anthropologists Marlize Lombard and Peter Gärdenfors agree with Povinelli. Their argument is interesting because it really unpacks what it means to say that an animal understands causality. They argue that to understand cause-and-effect fully means understanding an idea like "the wind *made* the apple fall from the tree." It seems so simple to us, doesn't it?

If you analyze this idea, however, you can see that it requires *if-and-then* systemizing: *if* an apple is attached to a tree, *and* the wind blows, *then* the apple will fall down. What's behind understanding this event is an understanding of the causal factors

Figure 6.2.
Chimpanzee causal systemizing tests.

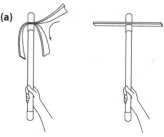

(a) The stiff rake would be much more effective than the floppy one to retrieve a food item.

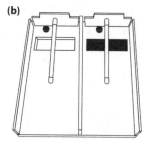

(b) The rake that would not cause food to fall into a trap would be more effective than the one that would cause it to fall into a trap.

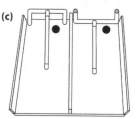

(c) The inverted rake would be more effective than the one that was the right way up with which to retrieve food.

within a system—that wind is a force, that fastening is a mechanism that can resist a force, but that if the fastening is insecure, the force of the wind could detach it. They conclude that we don't see apes and monkeys fastening one thing to another, to secure them, because there is no evidence that they understand such causal links. Yet we see people using bungee cords to fasten heavy loads to the roof bars on their car before they drive at speed on a motorway, or fastening down their tent using ropes and tent pegs driven securely into the ground before they go to sleep in the tent on a windy night. These mundane human activities reveal a set of causal concepts that we have and that other animals lack.[23]

Lombard and Gärdenfors also argue that hunting prey with a bow-and-arrow, let alone using a poison-tipped arrow, requires full causal understanding, and remind us that bow hunting is exclusive to humans and dates back to 71,000 years ago in South Africa.[24] This timing fits the theory that there was a cognitive revolution 70,000 to 100,000 years ago. As we discussed earlier, understanding a bow-and-arrow requires systemizing: "*if* I attach an arrow to a stretchy fiber, *and* release the tension in the fiber, *then* the arrow will fly." The key causal factors within this system are that for an object to travel any distance requires a force to act on it, and creating and releasing tension in a fiber is a way to control a force. Indeed, the very idea of any projectile (shooting an arrow, throwing a rock or a spear, or launching a cannonball, a bullet, or a rocket) requires an understanding of causality.

For me it is quite a shocking conclusion that non-human animals may not understand causality. From Lombard and Gärdenfors's analysis we should find that chimps don't throw spears. Is that really the case? Although chimps make spears to stab their prey with, their hunting with sharpened branches is still just very simple tool use. It turns out that spear-throwing may indeed be unique to humans. Chimpanzees may throw stones (another projectile) at zoo visitors to scare them away, but this certainly doesn't prove they understand causality—they could simply have learned the association between their action and an outcome.[25] Humans alone play darts and are supremely accurate at throwing things at a target, unlike apes or monkeys, who are remarkably clumsy. I love watching people in the pub playing darts, or watching dart champions on TV, not because it's always so exciting as a game, but more because that simple act of taking aim and refining the movement of the arm, hand, wrist, and finger-to-thumb grip betrays a beautifully elegant set of causal concepts. The same is true for kicking a football toward the goal or swinging a tennis racket or golf club to send a small ball toward a precise target.

So, tempting as it is to see the plethora of animal tool use as similar to our own tool use, none of these examples come close to the special invention capacity or causal reasoning capacity of humans, for two reasons: First, all of these examples are of *simple* tool use and could just be the product of associative learning.[26] And second, humans can generate an *infinite* number of new variations of tools, whereas most of these examples from non-humans are either a one-off instance or evidence only of a limited repertoire. And the absence of generative invention in any other animal strongly points to the evolution of the Systemizing Mechanism.

||||||

To give you a final sense of how complex tools like a bow-and-arrow differ from a simple tool like a stone ax, consider an argument from Indian doctor Saravanane Carounanidy. He points out that making a stone hand ax is just a two-step process: first, slice a stone along one side, then slice it down the other side. Anthropologists call this "flint-knapping." And it doesn't matter if you start on one side of the stone or the other. By calling it a two-step process, Carounanidy reminds us that the chances are very high that at some point an animal completed these two steps—by chance alone—and got the reward (some food).

So, the crow, by chance alone, might have learned that dropping a nut (A) in the path of a car (B) is associated with getting the food inside the shell. As we discussed earlier, associative learning that connects A with B could lead to this apparent invention. Without a genuine ability to experiment and invent, the animal would simply repeat the making of this simple tool, without modifying it. In contrast, Carounanidy argues, making a more complex composite tool—for example, a hunting spear with a flinthead—could not be accomplished in fewer than *six* steps, performed in a very specific sequence, such that they would be highly unlikely to occur by chance alone. Here are the six steps that would be required to make a spear with a flinthead:

1. Make a small ax from stone (itself a two-step process)
2. Find a long stick from the branch of a tree
3. Make a small cut (a notch) at one end of the stick
4. Find a natural fiber
5. Place the ax in the notch
6. Tie the ax firmly to the branch using the fiber

This six-step process could either be performed in this specific order, or where Step 1 is before Step 5, to produce a flinthead spear. The list of six steps contains an implicit "then" after each step. The number of possible permutations involved is therefore a huge 720 (or 6 factorial, which is $6 \times 5 \times 4 \times 3 \times 2 \times 1 = 720$). Therefore, the likelihood of making this incredibly useful new tool *by chance alone* is very low: 1 in 720. In other words, when *Homo sapiens* first made a flinthead spear, it was unlikely to have been by chance.

For *Homo sapiens* to create the flinthead spear, both a greater working memory capacity (the number of steps had jumped from two to six) and the new *if-and-then* algorithm were required to understand which steps must happen before which in the tool-making process. There is some speculation that

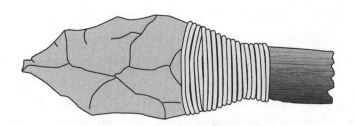

Figure 6.3. Wooden spear with stone ax tip

Neanderthals made spears, but whether they attached stone ax tips is still a matter of debate.[27]

In contrast, by 70,000 years ago *Homo sapiens* hadn't just invented the flinthead spear but, as I mentioned earlier, had also made the bow-and-arrow. Using Carounanidy's analysis, we can see that the construction of the bow-and-arrow needed a minimum of nine steps, each of which had to occur in one of two prescribed orders (as below, or where Steps 1–3 switch with Steps 4–7):

1. Take a long and flexible stick
2. Find a stretchy fiber or animal tendon
3. Tie the fiber tight to both ends of the stick, so that the stick bends into a bow shape
4. Take a short, straight stick
5. Make a small blade from stone (itself a two-step process)
6. Cut a small slit in one end of the short stick
7. Place the blade in the slit and tie it to the stick with a length of fine fiber
8. Place the bladed stick perpendicular to the bowed stick so the non-blade end is resting against the stretchy fiber
9. Pull back on the stretchy fiber and release the bladed stick

Again, the number of permutations is huge: 362,880 (or 9 factorial, which is $9 \times 8 \times 7 \times 6 \times 5 \times 4 \times 3 \times 2 \times 1 = 362,880$). So, the likelihood of the bow-and-arrow having been made by chance is vanishingly small: just 1 in 362,880—or near-impossible.[28] Rather, the reason humans 70,000 to 100,000 years ago were able to suddenly start making complex tools was because of the evolution of the Systemizing Mechanism.

Let's go back to the crow and the nut example. How can we rule out that the crow isn't thinking: "*if* I drop a nut on the road, *and* a car drives over it, *then* I can open the nut"? Couldn't it be that the crow is capable of *if-and-then* reasoning? In science it is argued that we should use the law of parsimony (also known as "Occam's razor" or the economy principle) to choose between two explanations: Can the behavior be explained more simply, without recourse to a more elaborate explanation? I would argue that associative learning is the simplest explanation for the crow's behavior, while the Systemizing Mechanism is the simplest explanation for runaway human invention.

||||||

The Systemizing Mechanism theory is a new theory of human invention, and of why other animals don't invent. But is there an alternative theory of human invention? To my mind, the main alternative theory is that humans started to invent because we alone evolved *language*. So far, we've hardly mentioned this enormous elephant in the room—language has ironically been silently sitting in the background. It's time to pit these two theories—systemizing versus language—against each other, head to head.

Chapter 7

The Battle of the Giants

Do we invent because we systemize, as I have argued, or because we have language? Which of these two powerful devices in the brain is the best explanation for human invention?

No one will deny that an obvious big difference between humans and other animals is that we have language and they don't. And this is not a new idea: the nineteenth-century linguist Friedrich Max Müller wrote that language is the "one great barrier between the brute and man," and Darwin, while acknowledging that other animals have forms of communication, discussed how they lacked the complexity of human language.[1] Contemporary writers have also highlighted the importance of language in enabling flexible thought. For example, paleoanthropologist Steve Mithen has argued that the inventions of art and sculpture 40,000 years ago reflected a shift from the modular mind of our hominid ancestors to the more integrated mind of modern humans, via language.[2]

It is easy to imagine that, once we evolved language, we could entertain hypothetical ideas of an *if-and-then* nature, enabling a capacity for invention.[3] It would be nice if the battle between these two big contenders as theories of invention could be settled by chronology—which came first, the Systemizing Mechanism or language? Sadly, this option is not available to us because the Systemizing Mechanism and language likely originated around the same time, around 70,000 to 100,000 years ago. So a different way to resolve this question is to use the principle of parsimony: Does invention need language, or can invention happen in the absence of language? Expressed differently, can the Systemizing Mechanism account for invention without needing language?

The word *language* is actually not that helpful, because it's an umbrella term. So, let's start by breaking language down into some of its key components, beginning with *speech*. We can rule out speech as being sufficient for the capacity for invention because the physiological apparatus for speech was present in our hominid ancestors at least 600,000 years ago. We know this because the hyoid bone in the front of the neck—thought to be key for speech and articulation—existed (in modern human form) in *Homo erectus* and also in Neanderthals.[4] In contrast, as we saw in chapter 5, generative invention only appears in the archaeological record from about 70,000 to 100,000 years ago. So, speech alone can't explain the flowering of human invention during the cognitive revolution.

What about *communication*, which is different to speech? Many species have communication systems even though they don't have speech. For example, bees do the "waggle" dance to signal to other bees where to find the pollen, and birds sing (often at dawn) to attract opposite-sex members of their species to find them in the mating season.[5] Vervet monkeys make alarm calls if they spot a tiger, a snake, or an eagle, causing other vervet monkeys to climb a tree, look at the grass, or look up at the sky, respectively, depending on which predator has been "announced."[6] And yet, as we saw in chapter 6, we don't see other species inventing. So, having a communication system per se also doesn't seem to be sufficient for the capacity to invent.

So, is there some other aspect of language that might have enabled humans to invent? What about *recursion,* which, according to linguist Noam Chomsky, is the unique feature of human language?[7] (Recursion is when a procedure includes the procedure itself, and which can repeat indefinitely.) I'm going to explore recursion in a bit more depth because it's so remarkable, and because it's a good contender for how we invent.

One example of recursion is "nesting." If you take the sentence "Alex has a red car" and nest within it "whom you know very well," you can make the sentence "Alex, whom you know very well, has a red car." The remarkable power of recursion is that we can keep nesting, building up more and more layers, like Russian dolls. So, the phrase "that is parked there" can be nested into our previous example, to become "Alex, whom you know very well, has a red car that is parked there." Such recursion can go on, in theory, indefinitely. One could easily imagine how, once humans had this capacity for taking phrases, nesting them inside others to build ever more complex linguistic structures, this could have been very useful for the more general capacity to invent new things. Essentially, these are effectively the building blocks of a sentence (the clauses and subclauses of grammar).

A second example of recursion is how, with a finite number of words, we can create an infinite number of sentences. Neurobiologist Andrey Vyshedskiy invites us to imagine having a language with 1,000 nouns, including "bowl" and "cup." He gives the example of adding the spatial preposition "behind" to our 1,000-word vocabulary. Suddenly we have a huge number of three-word phrases, such as "bowl behind cup," or "cup behind bowl." In fact, the number of distinct mental images we can refer to increases from 1,000 to 1 million (his calculation is $1,000 \times 1 \times 1,000$).

Now Vyshedskiy invites us to imagine adding a second spatial preposition, such as "on." We can now generate a huge number of five-word phrases, such as "cup on plate behind bowl." Suddenly we've got a lot more we can talk about than just pointing at a bowl and saying "bowl." In fact, he calculates, by adding these two spatial prepositions to our 1,000-word vocabulary, the number of distinct mental images we can refer to jumps to 4 billion (1,000 x 2 x 1,000 x 2 x 1,000). This is another remarkable example of recursion. Vyshedskiy calls this powerful increase toward an infinite number of sentences "magic."

So, is the recursive property of language a rival theory of human invention? I don't think so, for several reasons.

First, recursion is not just found in language—it's also a critical feature of music.[8] Given the key role of systemizing in the invention of music, as we discussed in chapter 5, this suggests that the Systemizing Mechanism likely enabled recursion, not the other way around. Consider how *if-and-then* reasoning could handle the earlier recursion example: "*if* I take the phrase 'Alex has a red car,' *and* nest within it 'whom you know very well,' *then* the phrase becomes 'Alex, whom you know very well, has a red car.'"

Secondly, people who lose language as a result of a stroke, or who never developed much language in the first place, can still be remarkable musicians.[9] This again suggests that you don't need linguistic recursion, but you do need the Systemizing Mechanism, to be able to invent music.[10]

Thirdly, a human mother can mesmerize her infant with varying rhythmic patterns during simple games like "Pat-a-cake, pat-a-cake, baker's man," long before her infant can handle linguistic recursion.[11] This suggests that infants can pick up *if-and-then* patterns without linguistic recursion.[12] Such rhythms can include nested ones, where different sequences can be nested within each other. For example, a mother could chant the "pat-a-cake" rhythm, then switch to the "incy-wincy-spider" rhythm, and then switch back to the "pat-a-cake" rhythm, and an infant would be able to follow the rhythm simply by using *if-and-then* reasoning.

Let's turn to a final key feature of human language, *syntax*. Syntax is remarkably powerful. It allows us to change the phrase "dog bites man" to "man bites dog," each with a very different meaning, just by breaking the phrase down into its individual units (three words, in this case) and then changing the sequence of the first and last words (the subject and the object). It is easy to imagine how syntax allowed the mind to create new images or ideas, which is essentially invention.[13]

But again, I don't see the capacity for linguistic syntax as a challenge to the Systemizing Mechanism theory of invention, because syntax is also a property of the Systemizing Mechanism. Consider how the Systemizing Mechanism runs the *if-and-then* algorithm: *if* a numerical sequence is 1-2-3, *and* the first and last digits are swapped around, *then* the numerical sequence is 3-2-1. Word sequences can be pushed through the same *if-and-then* pipeline: *if* the phrase is "dog bites man," *and* the first and third words are swapped around, *then* the phrase becomes "man bites dog." Nor do I see that syntax is essential for invention, but, according to my theory, *if-and-then* reasoning is a requirement. Without it, there would be no invention.

So, *if-and-then* thinking conferred on humans 70,000 to 100,000 years ago the capacity to rearrange variables within *any* system: "*if* I take a straight bladed tool, *and* change its shape to a curve, *then* it can become a fishing hook." The Systemizing Mechanism allowed—and still allows—powerful magic: infinite invention. As a side benefit, *if-and-then* thinking enabled us to perform operations like recursion and syntax. These in turn would have transformed a simple language into a complex one. Undoubtedly it was a two-way street, with language facilitating *if-and-then* reasoning by allowing us to put our new ideas into words, and then play with words to come up with new ideas.

But the very existence of autistic savants, some of whom have very minimal language but who are hyper-systemizers and who can invent, suggests that systemizing and language are independent of one another.[14] Two beautiful examples of such autistic savants are Nadia, an autistic girl who could draw horses from any perspective, when she had almost no language; and Stephen Wiltshire, who could draw buildings with remarkable accuracy and from any perspective, even as a child and when he had very limited language. (Both Nadia and Stephen eventually developed some language.)

In sum, in my view, language is powerful in its own right, but is not a rival explanation of human invention.[15]

▮▮▮▮▮

Four other psychological theories have been put forward to explain the capacity for invention, and I'm going to deal with them briefly.

The first is that we invent because we can *integrate two ideas* into *one new one*. Vyshedskiy argues that this is the role of the lateral prefrontal cortex in the brain. According to him, this is what enabled early humans 40,000 years ago to take two separate concepts ("man" and "lion," for example) and combine them into a synthesized concept ("lion-man") to make a sculpture of such a fictional entity. Vyshedskiy suggests that only the lateral prefrontal cortex can combine objects from memory into a novel mental image. He calls his alternative theory of human invention "prefrontal synthesis."[16]

However, the lateral prefrontal cortex is involved in a lot more than integrating two ideas into one new one. And this theory isn't really an alternative because integrating two ideas is just an operation within the Systemizing Mechanism (it's the *and* in *if-and-then*). Thus: "*if* I take (the idea of) the top half of a lion, *and* I attach it to the bottom half of (the idea of) a man, *then* I have (the idea of) a lion-man." The power of the

Systemizing Mechanism is that it can perform this *and any other operation* on input to produce an invention, whether the input is real, or is an idea, a word, a picture, or a model (such as sculptures representing objects), to produce fictional entities (like Spider Man).[17]

The second theory, briefly, is the idea that we became capable of invention because we could think symbolically.[18] Archaeologist April Nowell has proposed exactly this. One meaning of the term *symbolic* thinking is the capacity to let one thing stand for another, or to imagine that one thing stands for another, as in algebra when we say, "*If* x represents the number of apples in this box," or in drawing when we say, "*If* this big circle I have drawn in the sand represents the Earth." So, the first meaning of symbolic thinking involves *hypothetical* thinking.

This was undoubtedly a huge step forward in humans' cognitive power—to be able to entertain thoughts about hypothetical situations—and it's unclear if any other species is capable of this. But hypothetical thinking is not really a challenge to the Systemizing Mechanism theory of human invention because hypothetical thinking is the *if* element in *if-and-then* systems-thinking. In addition to hypothetical thinking, someone also had to use *if-and-then* thinking to see how ochre could be used as paint, or to make a tool like a paintbrush, or a chisel to carve a sculpture. Systems-thinking (*if-and-then* reasoning) had to come first.

A second use of the term "symbolic thinking" is what psychologist Alan Leslie calls the capacity for *meta-representation*. During meta-representation, a proposition ("the moon is made of blue cheese") is prefixed by a mental state (for example, "I *imagine* that..."). The result is a sentence: "I *imagine* that the moon is made of blue cheese."[19] This statement can be true even if the statement "the moon is made of blue cheese" is obviously false. Meta-representation enabled both theory of mind (imagining someone else's thoughts) and self-awareness (thinking about one's own thoughts), and it is one part of symbolic thinking (pretending or imagining that one thing represents another).

Meta-representation could be argued to be key to the capacity for human invention because it would enable the thought "I *imagine* that this hollow bone could be used to make sounds." Meta-representation therefore allows us to imagine fictional possibilities and can explain our capacity for "make-believe." Make-believe is fun (it allows us to joke around and pretend a banana is a telephone) and socially invaluable (it underpins our ability to have a theory of mind, so is part of the Empathy Circuit). Being able to think "I *imagine* that x" must again have been a huge step forward in cognitive terms, and we have no evidence that any other animal is capable of such a thought.

But by itself, meta-representation cannot explain the capacity to engineer a new product. You can imagine or joke around as much as you like, but actually engineering something to understand technical implementation, still requires *if-and-then* systems-thinking. This is not to diminish the huge importance of symbolic thinking—as part of the Empathy Circuit—for art, language, and thought, but symbolic thinking does not replace the need for systemizing in explaining human invention.

A third theory comes from Yuval Harari, who argues that human invention is possible because we are the only species that can think about *collective fictions* (like religion, a limited company, or money).[20] He is of course right to underline how unique humans are in this respect, and how powerful such collective fictions can be: collectively sharing the same fictional belief can coordinate the activity of thousands or millions of people. As I write this, I am in awe of how, by the last week of March 2020, virtually all 7.6 billion people on the planet, believing a lethal but invisible virus was all around us, stayed home for several months, leaving the whole planet deathly silent and devoid of people in public places.[21] Such is the power of a *collectively held belief* (in this case, not a fictional one, but based on hard evidence of its reality) to mobilize large numbers of people into coordinated action.

However, again, we can think about and share fictional concepts as much as we like, but without a Systemizing Mechanism, we aren't going to be able to implement our ideas at a technical level. Fiction-thinking, collective or otherwise, only gives us the *if* part of *if-and-then* reasoning. We also need the *and* part, typically a causal operation, and as we saw in chapter 6, non-human animals don't appear to understand causality. And we need the *then* component, which allows us to see the results of observing or experimenting with causal operations. There is no convincing evidence that other animals can systemize the whole *if-and-then* pipeline.

The final psychological theory is that we can invent because we have a bigger *working memory.*[22] Archaeologist Thomas Wynn and psychologist Frederick Coolidge, who proposed this theory, define working memory as the ability to hold something in mind in the face of distraction. They argue that for humans to have designed and used traps, for example, they must have needed working memory: you set the trap, and then you watch and wait, or come back later, to see if it worked. Ironically, although the word "memory" usually refers to past information, the term "working memory" is also used in relation to implementing a future plan. That's because you have to be able to remember the steps of the plan.

There's no doubt that the human capacity for holding many more steps in mind must be a big advantage, but did increased working memory per se lead to our capacity to invent? The answer must be no, for several reasons. First, an animal can have a good working memory but still lack the capacity to invent. For example, squirrels have an excellent working memory for where they buried their nuts before the winter, yet don't invent in any generative way. Some have claimed that crows and apes can even think about the future, although this is contested, yet they also don't invent in a generative way. So, invention entails more than just working memory.[23]

Making a trap, a good example of planning, clearly requires way more than working memory. At a minimum, it also needs systemizing: "*if* I attach a spring to a metal bar, *and* trigger the spring, *then* the metal bar will snap shut." Or, "*if* a mouse nibbles the cheese, *and* this triggers the spring, *then* the metal bar will snap shut on the mouse's head, killing her in a split-second." Making a trap is also a sign of the capacity for deception and so entails cognitive empathy, or theory of mind (part of the Empathy Circuit), in that theory of mind is needed to appreciate that the mouse won't *know* what's about to hit her, or won't appreciate how the spring mechanism works. But systemizing is needed to design the spring mechanism in the first place.

In summary, although these four psychological processes are contenders for explaining human invention, and undoubtedly helped the whole process of invention, none of them replaces the need for a Systemizing Mechanism, and none of them alone would lead to invention. Furthermore, part of what is needed is a theory not only of *how* we invent, but *why* we invent. Recall that Edison was inventing for the pure pleasure of inventing. He worked on many of his inventions, not to meet an unmet need, but just to see what happens and what's possible. The Systemizing Mechanism is what drives curiosity.

|||||

We also need to briefly consider some alternative theories of invention that focus on evolutionary changes in the human body. Some have proposed our *upright posture* and our *overall brain size* as contenders, but *Homo erectus* also walked upright, and Neanderthals had an even larger brain than ours, and yet neither of them came close to our capacity to invent generatively.[24]

Another theory focuses on humans' *opposable thumbs*, which allow for more precise fine motor control, including a precision grip and a power grip, undoubtedly advantageous in advanced tool use. (Think how you grip an overhead rail in the subway compared with how you manipulate chopsticks.) But opposable thumbs can't explain how we can invent, since *Homo habilis*, all Old World monkeys, and all the great apes have opposable thumbs and yet these other primates don't invent generatively.

Finally, others have argued that our *long childhood* must be relevant. Undoubtedly this had an impact on our learning capacity: a protracted childhood means that human infants are born at a relatively more immature stage of development, so a bigger fraction of our knowledge is the result of experience rather than genetic pre-programming, increasing our behavioral flexibility. But a longer childhood per se doesn't automatically lead to a capacity for invention.

A final challenge to the Systemizing Mechanism theory of human invention might come from archaeological evidence that challenges the date of the cognitive revolution: Are there not apparent inventions that predate the idea that the cognitive revolution occurred 70,000 to 100,000 years ago?[25] The evidence of burial, the existence of perforated shells, and the use of pigments go back hundreds of thousands of years and may be cases of invention that predate modern humans. However, archaeologists have argued that these may not meet the criterion for being genuine inventions, largely because they are one-off instances and are open to other interpretations.

So, for me, none of these alternative proposals—psychological or physical—are sufficient to explain our remarkable capacity for invention. If we just have the *if* (as in hypothetical thinking), that doesn't get us to invention. If we just have the *and* (as in the concept of causality), that too doesn't get us to invention. Equally, if we have *if-then*, that doesn't help us understand how to invent, only that objects or events can change. To invent, I argue, we needed the whole process of *if-and-then* reasoning. Invention just can't be done without the Systemizing Mechanism.

||||

I've made the argument for why the Systemizing Mechanism is necessary for invention, and for it being partly genetic. This means hyper-systemizing qualities can be passed down from parents to their children. I've also presented evidence that the genes for systemizing partly overlap with the genes for autism. Indeed, this genetic overlap was one of the intriguing connections that we glimpsed at the beginning of this book, between the minds of hyper-systemizers such as inventors and the minds of autistic people, both of whom are drawn to seek *if-and-then* patterns in the world. This leads to a very specific prediction: that hyper-systemizing parents are genetically more likely to have an autistic child. To test if this is true, we need to observe what happens when hyper-systemizers breed. And the perfect opportunity to do so is found in places like Silicon Valley, where hyper-systemizers flock to work, then meet and start to make babies. It's a natural experiment.

Chapter 8

Sex in the Valley

Are hyper-systemizing couples, like those talented individuals who work in STEM fields, more likely to have an autistic child? Given that we now live in a world where individuals with good technology skills move to work in technology hubs, where they are more likely to meet each other, could these couples be contributing to the rising rates of autism? Increasing autism rates of course reflect greater awareness of autism, better diagnosis, and changing diagnostic criteria, but is there more to it than this?[1] Answering this question is important both to test the central theory of this book—that the genes for autism drove the evolution of human invention—and to plan for how best to support autistic people.

To find out if hyper-systemizing couples are more likely to have an autistic child, we conducted the Parents' Occupations Study, way back in 1997. It was the first large-scale survey of the occupations of parents of autistic children.[2] We sent a questionnaire to 1,000 parents of autistic kids and to a control group, asking about the parents' and the grandparents' occupations. We predicted that fathers and grandfathers of autistic children would be more likely to work in the field of engineering, a clear example of a hyper-systemizing occupation. This was clearly confirmed: they were more than twice as likely to be engineers, compared to the fathers and grandfathers of kids with no diagnosis of autism or of kids with a different disability. And this was true of the grandfathers on both the father's and the mother's side. At that time, too few mothers were working outside the home for us to know what kinds of jobs they might have chosen, but a separate study found that mothers of autistic children score higher on tests of systemizing.[3] So, genes from hyper-systemizing parents and grandparents seemed to be contributing to the likelihood of autism in a child or grandchild.

The Parents' Occupations Study was retrospective: it started with a child's known autism diagnosis and then worked backwards to see if the parents were hyper-systemizers. But could we find any prospective evidence? If you start with hyper-systemizing couples, are they more likely to have an autistic child?

I meet couples who have an autistic child all the time. Take couples like Jim Simons, the gifted mathematician and hedge fund entrepreneur, and his wife Marilyn Hawrys, a quantitative economist, who have an autistic daughter. Like many parents of an autistic child, Jim has used his hyper-systemizing to great advantage. He hired mathematicians and computer scientists to create computer models to predict the behavior of the financial markets. Jim's personal wealth is estimated at over $15 billion.[4] Or take Steve Shirley, who arrived penniless in England on the *Kindertransport* from Nazi Germany, studied math, and together with her husband Derek, a physicist, also had an autistic son. She found that, by separating software from hardware in the early days of computers when these only came as a bundle, she could create a business that made her a multimillionaire.[5]

Such couples are consistent with the idea of a genetic link between hyper-systemizing talent in parents and a higher likelihood that they will have an autistic child. Anecdotally, of the 330 wealthiest families in the United States, 27 (8 percent) have an autistic child.[6] Given that autism is found in 1 to 2 percent of the general population, this suggests a fourfold increase in the rate of autism in families in which one or both parents have been hugely successful at making money. Making money is usually the result of running a business, and a business is a system, so a successful business man or woman is likely to be a strong systemizer.

To see how key *if-and-then* thinking is to business, consider this example: "*if* I make $50 selling one unit, *and* I scale up production to one million units, *then* I will make $50 million." A business, in taking something and turning it into a product for sale, is itself also a system, and if the *if-and-then* process is a new business pipeline, it would meet the criterion of being an invention.

So, a hyper-systemizing brain type, which is a massive asset when it comes to invention and wealth creation, could be linked to a greater likelihood of producing an autistic child. Anytime one comes across a highly successful entrepreneur—someone who has clearly systemized their business and who also, having the technical knowledge central to their business product, has therefore systemized their product—they might be expected to be more likely to have an autistic child or grandchild.

Of course, hyper-systemizing in a parent or grandparent may not just manifest as financial wealth derived from business acumen: it could also show itself as scientific, academic, technical, literary, or musical expertise. Consider the remarkably successful physicist Stephen Hawking, who has an autistic grandchild.[7] Or Elon Musk, perhaps the world's most famous innovator and inventor, who has an autistic child.[8] These anecdotes hint that hyper-systemizing grandparents and parents are more likely to have an autistic child or grandchild. But to move beyond anecdote to evidence, to test if this link is genetic and not due to chance, we need to look at rates of autism in a large population of hyper-systemizing parents.

||||||

I had been hearing anecdotal reports that autism was much more common in Silicon Valley since at least 1997. I had also read an article anecdotally suggesting that autism was more common among children of alumni of the Massachusetts Institute of Technology (MIT). If these rumors were backed up by data, they could confirm the genetic link between hyper-systemizing and autism.

And then, in 2003, I received an interesting email from the former president of the MIT Alumni Association, Brian Hughes, who told me that among his alumni, autism rates were anecdotally reported as high as 10 percent, not the usual 1 to 2 percent. If the rates of autism really were five-fold higher in this unusual population, this would be super-important. With Brian and psychologist Sally Wheelwright, we set up the MIT Study, a survey that would go out to thousands of MIT alumni.

MIT was an interesting place to choose to stage the largest experiment on the mating patterns of systemizers because, before 1975, MIT only admitted men to study at the university, and the only courses on offer were in the "exact" sciences (STEM). So, if these men later ended up having autistic children, we could assume that, at a minimum, those children had a father who was talented in STEM. In 1975, MIT went coeducational, giving us the opportunity to study couples who had met there after this date and who had started a family. These would be couples in which both parents were talented in STEM, so we could see if they were more likely to have an autistic child. We set out to compare the rates of autism among the children of couples in which both parents were in STEM, one parent was in STEM, and neither parent was in STEM.

(And of course there are lots of couples in the general population where neither parent is talented in STEM, effectively the control group.) The study received formal approval by the MIT Institutional Review Board, so we were good to go.[9]

And then another shocking email from Brian arrived in my inbox. He told me that the president of MIT, Charles Vest, had intervened to say that he wasn't willing to authorize this study to go ahead. Soon after, Vest retired but Brian still was told the study couldn't go ahead for fear that, if the hypothesis was confirmed, the reputation of MIT might be harmed. The study had been blocked, from the highest level in the university.

I was shocked for two reasons. First, scientifically, this was a really important question to answer, to help understand the link between autism and systemizing talent. And secondly, MIT's decision challenged the important governing principle in universities of academic freedom of ideas.[10] I have worked in universities for thirty-five years, and I have never ever heard of the president of a university—anywhere in the world—dictating what research could or couldn't be conducted within their university, unless the research contravened legislation or might have been morally offensive. Other than these kinds of reasonable limits on academic freedom, the principle is inviolable. Nevertheless, I had no wish to embarrass MIT, so I had to drop the plan. But I was not going to just abandon my attempt to test this important hypothesis. I would simply have to find a different location for the study.

Then one day in 2010 my luck changed. I received another email, this time from Patrick Wiercx, a Dutch journalist from the city of Eindhoven in the Netherlands. He was writing a story about the anecdotal reports of very high rates of autism in Eindhoven, the Silicon Valley of the Netherlands. I invited Patrick to come over to discuss this idea, and he jumped on a plane straightaway. He explained that Eindhoven was a hub for hyper-systemizers because of two big magnets that drew them: the Eindhoven Institute of Technology (another MIT) and the Philips factory, which has been there for over a hundred years. We decided to work together.

By good fortune, I had a Dutch master's student, Martine Roelfsema, who could help us navigate the cultural challenges of such a study. With a team of epidemiologists, we designed the Eindhoven Study to see how many autistic children there were in Eindhoven compared to two other Dutch cities, Utrecht and Haarlem, which had similar-sized populations and were comparable on other demographics, but which, unlike Eindhoven, were not STEM hubs.[11]

Martine contacted every primary and secondary school in all three cities—over 650 schools—to ask them how many children were already on their special educational needs register as having a diagnosis of autism. More than half of these schools agreed to take part, providing information on over 60,000 children. When the results came in, we were astonished—they exactly fitted our predictions. In Eindhoven, there were 229 cases of autism per 10,000 children, compared to just 84 in Haarlem and 57 in Utrecht. So, autism was more than twice as common in Eindhoven.

From this, we concluded that people talented in STEM are more likely than people in the wider population to have an autistic child, and in such communities we should see a spike in the prevalence of autism. This needs to be replicated in other STEM hubs like Silicon Valley itself and may have relevance for understanding the exponential rise in the rate of autism in the digital age, although of course many factors might contribute to this.[12] But the evidence from Eindhoven strongly suggests that inheriting a hyper-systemizing mind makes a child more likely to be autistic.

Let's return to the results of the Parents' Occupations Study, which showed that, in a family with an autistic child, both the maternal and paternal grandfathers were more likely to work in engineering. This presents a number of puzzles.

First, what attracts these men and women to each other in the first place? When they met, they didn't know they each had a parent who was more likely to be an engineer. And a related puzzle is this: How do two individuals carrying hyper-systemizing genes end up as a couple? If it is just physical attraction, why should they have similar minds? If it is mental attraction (called sapiophilia), again, why should they have similar minds? I was intrigued by what makes two hyper-systemizers more likely to end up as a couple.

"Like marrying like" is what biologists call "assortative mating."[13] Assortative mating is widespread in nature. For example, in humans, tall people are more likely to marry tall people, extraverts are more likely to marry extraverts, and alcoholics are more likely to marry alcoholics. Based on our Parents' Occupations Study, it seemed that hyper-systemizers were more likely to marry hyper-systemizers. This fits the "assortative mating" theory of autism, which predicts that autism rates will be higher among hyper-systemizing couples' children or grandchildren. Interestingly, this has recently been confirmed at the genetic level.[14]

There are a few ways in which assortative mating among hyper-systemizers might happen. It could be that people who share similar interests are more likely to end up living in the same location, because they have pursued comparable paths in their education and career choices. Or perhaps hyper-systemizers are more likely to become a couple because more socially skilled individuals form couples sooner, and less socially skilled individuals are the ones who, years later, are still available and looking for a mate.[15] They find each other because they're both still single. A third possibility is that assortative mating among hyper-systemizers might be the result of two people being attracted to each other because their *minds* are similar.[16] But how could we test this?

We invited mothers and fathers of autistic children to take a systemizing test. We used the Embedded Figures Test, a pattern recognition test that involves finding a target piece hidden within a design. We had previously given this test to autistic adults, who are faster and more accurate at it than are typical adults, probably because, being hyper-systemizers, they break down information into its component parts as rapidly as possible, to then look for *if-and-then* patterns.

Sure enough, we found that both the mothers and the fathers of autistic children were also faster and more accurate on this pattern recognition test.[17] They shared a skill with each

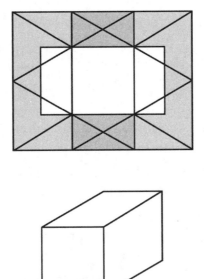

Figure 8.1. The Embedded Figures Test. Autistic people and their parents are on average faster than typical people at finding the target (the cube below) hidden in the design (above).

other, related to hyper-systemizing, that most people don't have. Of course, they were totally unaware of their shared skill on this test when they began dating and later started a family, but one possibility is that each was attracted to the other's detail-oriented, laser-sharp mind, or at least wasn't deterred by it. Perhaps they felt comfortable with a person whose mind was just like theirs, and just like the mind of one of their parents.

||||||

Eindhoven was a "natural" experiment that allowed us to observe what happens when you speed up evolution to see how the children of hyper-systemizers develop. And the Eindhoven Study again provides evidence consistent with the idea that the genes for systemizing overlap with the genes for autism. We can therefore assume that the hyper-systemizers who drove the evolution of human invention, starting 70,000 to 100,000 years ago, themselves had a high level of autistic traits and were more likely to have a child with a high number of autistic traits. We can assume this because it is still the case today. But back then, the likelihood of two hyper-systemizers having children was much lower. Today, as more Eindhovens and Silicon Valleys take root and blossom in every country, bringing hyper-systemizers together who make families, what might we expect the future to hold?

Chapter 9

Nurturing the Inventors
of the Future

Eindhovens and Silicon Valleys are springing up across the planet—from Silicon Alley in New York to Silicon Roundabout in London, Silicon Fen in Cambridge, Cyberabad in Hyderabad, and the Silicon Valley of India, Bangalore.

This means that, as more and more people who are wired for hyper-systemizing are meeting and having children, we can expect more autistic children to be born in these communities. We need to anticipate and plan for the special needs of these autistic children who may require—and who have a right to—lifelong support. Some will have additional intellectual disability, but at least half of them will have average or above-average IQ.[1] If we want to nurture the inventors of the future, the next Thomas Edison or the next Elon Musk, we are more likely to find them among autistic people, and among those who have a high number of autistic traits because they are hyper-systemizers, than among the general population.

The minds of autistic people who have no intellectual disability and who are hyper-systemizers should be seen as one of many natural types of brains that have evolved and that add to human neurodiversity. Autistic people and hyper-systemizers without a diagnosis represent just one type of brain among many, and may excel or struggle, depending on what environment they are in. As one Danish autistic man told me:

> "We are like freshwater fish in saltwater. Put us in saltwater and we flail around, suffer, struggle, and may even die. Put us in freshwater and we flourish."

Psychologists used to talk about "normal" children and described all other types of children as "abnormal," as if there were just one kind of normal brain. The concept of neurodiversity is a refreshingly different view of the range of brain types out there, because it acknowledges that there is no single way for the brain to develop.[2] I would call neurodiversity a "revolutionary" concept because it introduces a radical new view of the world—one with many varieties of brains, all of which occur naturally, rather than the old, inaccurate, binary view of normality vs. abnormality.

The neurodiversity view is that there are diverse pathways in development: some people are more verbal, some more spatial, some more musical, some more mathematical, and some more social. All of these different brain types exist in the population, and this is by no means an exhaustive list of brain types. Consider how some children are right-handed, some left-handed, and some ambidextrous, since handedness reflects differences in wiring in the brain. And some children have more sensory sensitivity, some more physical coordination skill, and some are color-blind. The list goes on.

Neurodiversity is simply a fact, just as biodiversity is a fact. Some estimate that up to 25 percent of the general population are "neurodiverse," if we include a range of disabilities, from autism to attention deficit hyperactivity disorder (ADHD) to dyslexia and dyspraxia.³ Another view is that we are all neurodiverse, in that we are all different. The five brain types we discovered in the population, defined by where individuals are on the empathy and systemizing bell curves, is consistent with that view. Different brain types likely evolved to be better adapted to certain environmental niches. Einstein reportedly said: "Everybody is a genius, but if you judge a fish by its ability to climb a tree, it will live its whole life believing it is stupid."⁴ The point is well made. Each of us should be judged for what we can do, not for what we can't do.

There are some who object to the notion of neurodiversity and its application to conditions like autism, which they prefer to argue is a disease and therefore needs a cure. My view is that when it comes to autism, all the four Ds apply: difference, disability, disorder, and disease.⁵

Difference in those with autistic minds is obvious: a freshwater fish and a saltwater fish are neither normal nor abnormal— they've each evolved differently in order to function and blossom in a particular environment but will struggle in another. Difference may be physical, as in some people having blue eyes and others having brown, or being taller or shorter. Differences may also be psychological, with different people having different profiles of strengths and weaknesses. Think of those who have an exceptional memory or who show remarkable attention to detail, or those who are less comfortable with casual chatting and prefer structured activities.

Disability is where an individual is below average in a particular skill, or has a condition that affects their ability to function in everyday life, and so they need support. An example might be a child who is not yet talking by five years old and needs help to communicate. *Disorder* is where an individual suffers from one or more aspects of their difference, the cause of which is not yet known. An example might be gastrointestinal pain, which, for unknown reasons, occurs in many autistic people.

And a *disease* is where that "symptom" is causing suffering and where the cause is known. An example might be epilepsy, which occurs in a proportion of autistic people.

The first two of these four Ds, difference and disability, are entirely compatible with the notion of neurodiversity. Some view neurodiversity as nature's strategy for ensuring human minds can cope with the diversity of challenges that different environments present. Harvey Bloom, writing in an early article on this topic, explored the idea that neurodiversity might be as important in evolutionary terms as diversity in the genome or the biome: "Neurodiversity may be every bit as crucial for the human race as biodiversity is for life in general. Who can say what form of wiring will prove best at any given moment?"[6]

Autism involves hyper-systemizing, which leads to strengths, and it also gives rise to disabilities—difficulties in communicating and socializing and in reacting to unexpected change. But in the right environment, such disabilities can be minimized. In this way, the disabilities are largely a product of the *fit* between the individual and their environment. When the hyper-systemizing qualities of autism are supported and nurtured, the unique skills and talents of autistic individuals can shine—to their benefit, and to the benefit of society.

Specialisterne is a Danish company that set out to create that right environment to allow the skills in autistic people to flourish in the workplace. The company was founded by Thorkil Sonne, technical director of a telecommunications firm, who set up this revolutionary company that hires only autistic people. A hyper-systemizer himself, he had watched with wonder as his autistic son displayed a talent at recalling train schedules and maps and building complex architecture and robots from Lego kits. Thorkil recognized that his son's ability would fit in well in a technically minded business environment.

But many autistic people wouldn't get the chance to demonstrate what they could do if they had to go through the standard job interview, which inadvertently discriminates against autistic people. Many of them prefer not to make eye contact or find it painfully stressful because "reading" another person's eyes is so confusing for them. Many of them may also struggle to communicate using words: they find it hard to read between the lines to pick up on their listener's hints or to know if they have given their listener too little information to make their communication clear, or too much information, which may bore or confuse their listener. All these communication skills require cognitive empathy, which autistic people find challenging. The result is that many autistic people experience high levels of social anxiety. So why ask them to go through a standard job interview, which might just be setting them up to fail?

Thorkil had the brilliant idea to develop an autism-friendly interview format. He asked autistic applicants to build and program Lego robots, for example, so that they could show their extraordinary pattern-seeking and problem-solving skills to potential employers. They could shine when given the task of looking for *if-and-then* patterns. Since then, a rapidly growing list of other companies have followed Specialisterne's example in an effort to increase the employment opportunities available to autistic people—one survey estimated that only 16 percent of autistic adults are in full-time employment[7]—and to realize the advantages to be gained from having employees on their team who think differently.[8] These efforts can be seen as part of a company's corporate social responsibility, since employing an autistic person is likely to improve that person's mental health by allowing them to feel included and valued and by reducing their social isolation.

One such company, Auticon (the name is shorthand for "autistic consultants"), only hires autistic people, who are offered jobs for life. They are placed as consultants in coding and other tech industry companies, whose autistic employees receive continuous support from Auticon on how to handle social interactions in the workplace. Auticon now has offices across Europe and in the United States. Another example is SAP, the multinational software development company that, thanks to V. R. Ferose, has developed an "Autism at Work" program. SAP aspires to have 1 percent of its 70,000-person workforce—700 employees—to be made up of individuals with a formal autism diagnosis. Ferose himself has an autistic child and ran SAP in Bangalore before moving to SAP's offices in Silicon Valley. Inspired by these two pioneering companies, many high-profile companies have since followed in their footsteps by launching similar neurodiversity employment programs.

Auticon says that autistic minds essentially use "a different operating system," a metaphor that author Steve Silberman also uses.[9] I like this metaphor because it draws attention to the fact that autistic people and other hyper-systemizers think differently: they are more objective, fact-oriented, and exact, and less subjective, emotion-oriented, or willing to approximate. One operating system isn't better or worse than another, but simply different, and designed to do different things. If you let a different operating system do what it's designed to do, it runs just fine. If you force it to do what it's not designed to do, it may crash and appear unable to function. Under the right conditions, hyper-systemizing can manifest as remarkable strengths and talents. Unstoppable *if-and-then* pattern-seeking works best in environments that do not change unpredictably. No wonder autistic people often find change so difficult that they resist it at all costs, attempting to live in a world that is as far as possible under their own control.

An employee's total focus, based on *if-and-then* patterns, brings rewards to teams and clients alike. Tom Monte, an SAP manager, says that the contributions of one of his autistic colleagues are invaluable:

"He uncovers things that we might overlook. He'll ask questions, and I think, 'Why aren't my senior guys asking these questions?'"[10]

HP Enterprise's neurodiverse teams are 30 percent more productive in software testing and debugging than teams of "neurotypical" employees.[11] Key to the success of these programs is the training and support given to autistic employees to ensure that they can manage the social aspects of the workplace—enough support that employees, such as SAP's Carrie Hall, feel confident to proudly identify themselves as autistic to their colleagues.[12]

The idea that autistic people think in a refreshingly different way as a result of their *if-and-then* pattern-seeking was beautifully borne out in a study of creativity and lateral thinking that used a classic test: "How many uses can you think of for a brick and a paper clip?"[13] Most people give predictable "simple" answers, such as using a paper clip to reset their iPhone. In contrast, most autistic people gave unexpected, less predictable, more "complex" answers, such as using a paper clip as a weight for the front of a paper airplane or heating up a paper clip to suture a wound. These answers made logical and scientific sense, but if you're neurotypical, would you have thought of such left-field examples?

The Israeli army has a special unit, called Unit 9900, for autistic people who wish to go through military service and whose excellent attention to detail and pattern-seeking talents can be harnessed to the army's needs.[14] Autistic people are asked to detect anomalies in satellite images of locations on Earth, to spot any patterns that look unusual. To some of us, this might seem like tedious work, but for a hyper-systemizer, it's "like a hobby," as one of the young autistic soldiers said. And the autistic Israelis in Unit 9900 are saving lives, spotting suspicious objects or movements. *If* it's an unusual shape, color, or movement, *and* it's different from its surroundings, *then* it could be terror-related.

The organization Ro'im Rachok now goes into schools in Israel to recruit autistic teenagers into this army unit so that they will feel equal to their peer group, experience a sense of inclusion, and feel valued for what they can bring. In societies where there is no conscription into the army, there are many jobs, such as looking at X-ray data in airport security or in hospitals, where it pays not to miss a single instance of an anomaly, such as a hidden weapon or a tumor. Indeed, one study confirmed that autistic security employees spot more suspicious items in X-rayed hand luggage.

James Neely, who is autistic, had difficulty keeping a job before he applied to Auticon. His sensory issues and social difficulties meant he needed to work in a quiet environment, and he now wears headphones to block out the noise while he is programming. His job-for-life at Auticon allowed him to recover from a long period of poor mental health, including depression, the result of trying to fit into workplaces that didn't understand him or offer him any help. In an interview with the *Guardian*, Neely reported that he is now happier, describing his placement with the pharmaceutical company GSK (GlaxoSmithKline) as "playing with data, just to see what we can do with it."[15] It's a win-win, benefiting both the company and its autistic employees' mental well-being.

We must scale up supported employment, not just for the benefit it will bring to society but because employment greatly improves the mental health of autistic people. Employment for autistic adults may turn out to be a far more effective intervention than any medical treatment, because it confers dignity and a sense of inclusion.

Of course, some autistic people don't express hyper-systemizing as a talent for computer programming or by being a modern-day Linnaeus. Instead they hyper-systemize by watching the washing machines going around all day, or lining up toys in strict patterns, or spinning objects. Many autistic people become lost in the detail and can't see the bigger picture. Yet, by looking at their autism as a form of hyper-systemizing, we might be better able to understand their world and to provide more opportunities for them to flourish at the same time.

Some parents will rightly protest that their autistic son or daughter, or even the majority of autistic people, do not have *Rain Man*–like "savant" skills, and that the day-to-day reality of autism is about multiple disabilities, diseases, and disorders.[16] Some of these parents and indeed some autistic people themselves are part of a vocal lobby group on the internet who are "anti-neurodiversity."[17] They may be taking this stance because they live with an autistic person who is clearly suffering, and consequently the whole family suffers too. These concerns are important, as there are indeed diseases and disorders associated with autism, including gastrointestinal pain, epilepsy, severe anxiety, severe learning disability, minimal verbal ability, and self-injury.[18] These can rightly be described as diseases and disorders in that they cause suffering. We must not ignore these challenging aspects of autism, and I share the wish for treatments or cures for such unwanted symptoms.

However, I would say, without wishing to cause offense to autistic people or their families, that these are not core features of autism. They co-occur with autism in some individuals, but by definition they are not core because they are not universal to all autistic people; they are not diagnostic of autism. It is absolutely right to demand treatments for unwanted symptoms, since ethically we should do anything to alleviate another person's suffering. But these symptoms do not characterize the autistic mind. Pattern-seeking and hyper-systemizing don't need to be cured any more than eye color does, which is also partly a reflection of a person's genetic makeup.

Returning to an example of disability, why do some 25 percent of autistic people have a major learning disability? This can arise from rare genetic mutations, such as in the *NRXN1* or *SHANK3* genes, which affect the development, structure, and functioning of the brain.[19] About one hundred of these rare genetic variants—which occur in only about 5 percent of autistic people—have been identified so far. Other factors that can increase the likelihood of learning disability co-occurring with autism include prematurity and birth complications.[20] For the majority of autistic people with learning disability, however, we don't yet know why their autism co-occurred with their learning disability.

One new hypothesis might be that a Systemizing Mechanism tuned high can produce a talented mind, but if tuned even higher, it manifests as learning disability. We can imagine that a person so exclusively focused on a tiny slice of data (the look and feel of sand grains as they pass through their fingers, or the shape and color of tiny soap bubbles in the kitchen sink) might have an extreme hyper-focus on detailed *if-and-then* patterns that interrupts their learning about the wider world, and even their language learning. But this is currently just a speculation awaiting research.

Alongside supported employment, autistic people should also be offered more support in negotiating the complexities of the social world, since the downside of hyper-systemizing is that it increases the likelihood of a disability in cognitive empathy. Interventions are available, and some of these help autistic people by harnessing their hyper-systemizing talents to learn social skills.

One example of such an intervention is *Lego Therapy*, where autistic kids enjoy and develop peer relationships in the safety of a systemizing activity they feel confident in, like building Lego models.[21] They learn to socialize and communicate in a context where they can use their *if-and-then* reasoning. And this is true of many people, such as those who enjoy socializing in a pub by playing darts or snooker, but who struggle to have conversation without a structured activity.

Another intervention for autistic children is *The Transporters*, an animation whose characters are all vehicles with human emotions appearing on their "faces," which are presented in the context of a highly systemizable domain.[22] These children can enjoy the predictability of trains, trams, and cable cars in these TV animations while also looking at faces and learning what situations cause them to change. A final example is *Mind Reading*, a digital resource that has systemized every human emotion, as expressed by actors.[23] *Mind Reading*'s encyclopedic video and audio format enables a person to learn to recognize emotions much as they might learn a foreign language: *if* a person's eyes are in shape A, *and* their mouth is in shape B, *then* their emotion is X. Each of these interventions has been evaluated and shown to lead to improvements in autistic children's social or emotion recognition skills.

Chris Worley noticed that her five-year-old autistic son Sasha naturally gravitated toward skateboarding. Sasha was good at it, probably because he had systemized all the moves with his *if-and-then* pattern-seeking skills.[24] His mother had the brilliant idea to start a foundation called ASkate. The idea was to use what Sasha was good at in order to overcome what he struggled with: learning to socialize through doing tricks with other skateboarders. Chris's foundation is a terrific example of taking a child with a disability and focusing on their talent. Chris set up ASkate in the car park of her local church in Alabama. Sasha loved skateboarding because he could experience total control over his board, and his mom realized this was how Sasha could be happy and be accepted as just another kid.

Ron Suskind's autistic son Owen had a different passion: Disney movies. Each sequence of the movies contains multiple *if-and-then* patterns that are evident if you watch them on a loop. Characters do and say the identical same thing every time.[25] Owen watched each movie thousands of times, until he knew every lyric and every character's intonation perfectly and could imitate them with exquisite accuracy—and yet he didn't talk to people or seem to understand them.

Ron had the remarkable insight that if he imitated different Disney characters, using the highly scripted voices in the exact intonations heard in the movies, this would not only grab Owen's attention but provide a scaffold and predictable structure so that Owen would provide the next phrase or next line in the dialogue—all from that specific movie. Ron would then ever so gradually adapt the "scripted" (that is, systemized) interaction to the current context, so that Owen would now not just be saying a line in the movie but talking about something in the current context. In this way, Owen started to communicate with Ron, and over many years he started to talk to others. Disney movies had become his stepping-stone into the previously confusing world of human social interaction and communication.

Anxiety is very common in autistic people—perhaps as many as 80 percent suffer from it. Common forms it takes are anxiety due to either unexpected change or social interaction. Is anxiety in autistic people so high because of the limits they encounter on which aspects of the world can be systemized? We discussed earlier how many autistic people may avoid the social world because many forms of social interaction can't be systemized.

Another new hypothesis that needs to be tested is whether, in an autistic person with preexisting anxiety and in whom the Systemizing Mechanism is tuned high, this might cause obsessive compulsive disorder (OCD).[26] OCD occurs at higher rates in autistic people than in the general population, and often takes an *if-and-then* form: "*if* there are germs on my hands, *and* I don't wash my hands through a strict sequence of actions, *then* I will contaminate others." Or, "*if* I contaminate another person, *and* that person dies, *then* it will all be my fault." While a Systemizing Mechanism tuned high might be highly advantageous for spotting patterns that help a person understand how something works, it may be that in the context of underlying anxiety it can also produce a disabling psychiatric degree of OCD. Again, this explanation for the elevated rates of OCD in autistic people is currently just speculation and needs testing.

How can we bring hyper-systemizing into education? Hyper-systemizers, including autistic people, learn differently, and while some will gravitate toward systemizing subjects such as math, physics, or music, others may even fail because many high school subjects are taught in a way that is not well designed for their kind of mind. These students will struggle in classes that are taught superficially or imprecisely, by a teacher focused on keeping most kids' attention at the expense of presenting factual information. Even worse, some school subjects are built around fuzzy tasks like writing a story rather than around rule-based systems like understanding how things work. Our schools should be identifying children from the earliest age as hyper-systemizers (which will include some children who have been formally diagnosed as autistic), so we can provide an educational environment that plays to their strengths, presenting information in an *if-and-then* format, so they can shine rather than fail or be turned off education altogether.[27]

As an aside, a hyper-systemizer only needs a diagnosis of autism if they are struggling to cope, to the detriment of their daily functioning. If someone has a supportive parent or partner to help them function, they may not need a diagnosis. Or if their lifestyle fits well with the characteristics of autism (such as being self-employed or having self-directed employment, together with a very accepting, tolerant, and open-minded set of neighbors or colleagues), again, they may not need a diagnosis. A diagnosis should be restricted to those who are struggling as a result of their autism.

Imagine an educational system that offered two streams: a broad curriculum—as we have now—for those who are generalists, which is most kids, and a narrow curriculum for those who are specialists, the hyper-systemizers. We have the tools to identify these kids because they are Extreme Type S on the systemizing and empathy bell curves. The broad curriculum already exists—this is what defines mainstream education. The focus is on learning a little about many subjects. But the broad curriculum doesn't work for some kids, because it involves too much switching too frequently. The broad curriculum is also often structured as group learning from a teacher, but some pupils learn better studying one-to-one, or even alone. The narrow curriculum would encourage those kids interested in one subject to go into as much depth as they want. It would be based on the idea that the most important thing is that the child has chosen a subject that ignites their passion and interest. Whether it's math or history or something far more focused, like an extinct ancient language, if the child simply wants to study only that subject for their whole semester, or their whole school career, it would still constitute a valuable education and prepare them for a specialist occupation. They should be allowed to pursue their strong narrow interest, sometimes pejoratively called their "obsession." Greta Thunberg, the Swedish autistic teenager, has a strong narrow interest in climate breakdown, and she has succeeded in raising awareness of the urgency of this issue for the future of the planet.[28]

I've met such people, and they blossom when given the opportunity. Daniel Lightwing, whom I diagnosed with Asperger syndrome when he was a student at my college, Trinity, in Cambridge, represented the United Kingdom in the International Math Olympiad. The documentary film *Beautiful Young Minds* is based on his life.[29] This is what he told me:

> "When I was about ten or eleven, it struck me that all of the subjects except maths and science are only relevant to, kind of, our civilization on this planet. But then maths, it studies everything that exists and everything that doesn't exist. And then, you know, from that moment I set apart everything else and just went to study maths."

For the kids who are natural specialists, a narrow curriculum designed for their strengths would prevent education from being a time of misery, which often results in underperforming and leaving school with few qualifications. Instead, they could enjoy their learning experience. We would be providing the right conditions in which their particular learning style could take root and blossom. Although some may argue that education should be broad, not narrow, if broad education is turning kids away from pursuing any education at all, a narrow education can surely be defended as better than none. As one parent put it to me: "Mainstream schooling is not fit for purpose for a proportion of children." Ironically, the generalist inevitably ends up specializing (by the time they get to college or into a job), and the specialist often discovers interesting connections between the initial highly focused starting point of their education and neighboring fields. So, these are just different routes into learning: broad to narrow for the majority, or narrow to broad for the tiny minority.

▌▐▌▐▌

Al, who we now know was Thomas Edison, had a high number of autistic traits, and Jonah is diagnosed as autistic. They are just two of the millions of hyper-systemizers who have driven human invention, and therefore human progress, over the past 100,000 years. Their minds are wired to seek patterns and to systemize insatiably, through eagle-eyed observation and careful, step-by-step experimenting. Among the new generation of hyper-systemizers will be some of the great inventors of our future. Their novel ideas can become innovations, but only with our support. If we acknowledge that some autistic people were and still are the drivers of the evolution of science, technology, art, and other forms of invention, their future can be different—but only with a big shift in our culture and society.

Appendix 1

*Take the SQ and the EQ
to find out your brain type*

TAKE THE SQ

The Systemizing-Quotient-Revised (ten-item version), or SQ-R-10

		Strongly agree	Slightly agree	Slightly disagree	Strongly disagree
1.	When I learn about a new category, I like to go into detail to understand the small differences between different members of that category.	2	1	0	0
2.	When I'm in a plane, I do not think about the aerodynamics.	0	0	1	2
3.	I am interested in knowing the path a river takes from its course to the sea.	2	1	0	0
4.	When traveling by train, I often wonder exactly how the rail networks are coordinated.	2	1	0	0
5.	When I hear the weather forecast, I am not very interested in the meteorological patterns.	0	0	1	2
6.	I enjoy looking through catalogs of products to see the details of each product and how it compares to others.	2	1	0	0
7.	When I look at a mountain, I think about how precisely it was formed.	2	1	0	0
8.	When I look at a piece of furniture, I do not notice the details of how it was constructed.	0	0	1	2
9.	When I learn a language, I become intrigued by its grammatical rules.	2	1	0	0
10.	When I listen to a piece of music, I always notice the way it's structured.	2	1	0	0

How to score the SQ-R-10:

For items 1, 3, 4, 6, 7, 9, and 10, if you strongly agree, you get two points for each; if you slightly agree, you get one point for each; and otherwise you get zero points.

For items 2, 5, and 8, if you strongly disagree, you get two points for each; if you slightly agree, you get one point for each; and otherwise you get zero points.

The maximum score is therefore 20, and the minimum is 0. The average score for females is 5.5 (the range is 2 to 9) and the average score for males is 6.7 (the range is 3 to 11). If you score 0 to 3, you are low in your systemizing drive, and if you score 12 to 20, you are high in your systemizing drive.

TAKE THE EQ

The Empathy Quotient (ten-item version), or EQ-10

		Strongly agree	Slightly agree	Slightly disagree	Strongly disagree
1.	I am good at predicting how someone will feel.	2	1	0	0
2.	Other people tell me I am good at understanding how they are feeling and what they are thinking.	2	1	0	0
3.	It is hard for me to see why some things upset people so much.	0	0	1	2
4.	I can easily work out what another person might want to talk about.	2	1	0	0
5.	I can't always see why someone should have felt offended by a remark.	0	0	1	2
6.	I can tune into how someone feels rapidly and intuitively.	2	1	0	0
7.	Other people often say that I am insensitive, though I don't always see why.	0	0	1	2
8.	In a conversation, I tend to focus on my own thoughts rather than on what my listener might be thinking.	0	0	1	2
9.	Friends usually talk to me about their problems as they say that I am very understanding.	2	1	0	0
10.	I find it hard to know what to do in a social situation.	0	0	1	2

How to score the EQ-10:

For items 1, 2, 4, 6, and 9, if you strongly agree, you get two points for each; if you slightly agree, you get one point for each; and otherwise you get zero points.

For items 3, 5, 7, 8, and 10, if you strongly disagree, you get two points for each, if you slightly agree, you get one point for each; and otherwise you get zero points.

The maximum score is therefore 20, and minimum is 0. The average score for females is 10.8 (the range is 6 to 16), and the average score for males is 8.9 (the range is 4 to 14). If you score 0 to 4, you are low in empathy, and if you score 16 to 20, you are high in empathy.

CALCULATING YOUR BRAIN TYPE

If you have completed both the EQ-10 and SQ-R-10, you can find out your brain type from the difference (D) between your SQ-R-10 score and your EQ-10 score (D = SQ-R-10 minus EQ-10). Your D-score can range from −20 to 20.

How to interpret your D-score:

Plot your score on the EQ-10 and the SQ-R-10 on the table below and look up your D-score. You can see your brain type either by color or by using these numbers:

Extreme Type E: −14 and below
Type E: −13 to −7 (inclusive)
Type B: −6 to −2 (inclusive)
Type S: −1 to 8 (inclusive)
Extreme Type S: 9 and above

For those who are statistically minded, the mean for the EQ-10 = 10.1, and the standard deviation = 4.9. The mean for the SQ-R-10 = 5.9, and the standard deviation = 4.0.[1]

SQ-R-10 \ EQ-10	1	2	3	4	5	6	7	8	9	10	11	12	13	14	15	16	17	18	19	20
20	-19	-18	-17	-16	-15	-14	-13	-12	-11	-10	-9	-8	-7	-6	-5	-4	-3	-2	-1	0
19	-18	-17	-16	-15	-14	-13	-12	-11	-10	-9	-8	-7	-6	-5	-4	-3	-2	-1	0	1
18	-17	-16	-15	-14	-13	-12	-11	-10	-9	-8	-7	-6	-5	-4	-3	-2	-1	0	1	2
17	-16	-15	-14	-13	-12	-11	-10	-9	-8	-7	-6	-5	-4	-3	-2	-1	0	1	2	3
16	-15	-14	-13	-12	-11	-10	-9	-8	-7	-6	-5	-4	-3	-2	-1	0	1	2	3	4
15	-14	-13	-12	-11	-10	-9	-8	-7	-6	-5	-4	-3	-2	-1	0	1	2	3	4	5
14	-13	-12	-11	-10	-9	-8	-7	-6	-5	-4	-3	-2	-1	0	1	2	3	4	5	6
13	-12	-11	-10	-9	-8	-7	-6	-5	-4	-3	-2	-1	0	1	2	3	4	5	6	7
12	-11	-10	-9	-8	-7	-6	-5	-4	-3	-2	-1	0	1	2	3	4	5	6	7	8
11	-10	-9	-8	-7	-6	-5	-4	-3	-2	-1	0	1	2	3	4	5	6	7	8	9
10	-9	-8	-7	-6	-5	-4	-3	-2	-1	0	1	2	3	4	5	6	7	8	9	10
9	-8	-7	-6	-5	-4	-3	-2	-1	0	1	2	3	4	5	6	7	8	9	10	11
8	-7	-6	-5	-4	-3	-2	-1	0	1	2	3	4	5	6	7	8	9	10	11	12
7	-6	-5	-4	-3	-2	-1	0	1	2	3	4	5	6	7	8	9	10	11	12	13
6	-5	-4	-3	-2	-1	0	1	2	3	4	5	6	7	8	9	10	11	12	13	14
5	-4	-3	-2	-1	0	1	2	3	4	5	6	7	8	9	10	11	12	13	14	15
4	-3	-2	-1	0	1	2	3	4	5	6	7	8	9	10	11	12	13	14	15	16
3	-2	-1	0	1	2	3	4	5	6	7	8	9	10	11	12	13	14	15	16	17
2	-1	0	1	2	3	4	5	6	7	8	9	10	11	12	13	14	15	16	17	18
1	0	1	2	3	4	5	6	7	8	9	10	11	12	13	14	15	16	17	18	19

EQ-10 Score

(From upper left to lower right in the table)

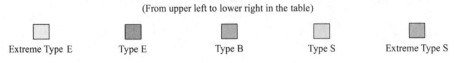

Extreme Type E   Type E   Type B   Type S   Extreme Type S

Appendix 2

*Take the AQ to find out how many
autistic traits you have*

The Autism Spectrum Quotient (AQ-10)

	Please tick one option per question only:	Definitely agree	Slightly agree	Slightly disagree	Definitely disagree
1.	I often notice small sounds when others do not.	1	1	0	0
2.	I usually concentrate more on the whole picture, rather than the small details.	0	0	1	1
3.	I find it easy to do more than one thing at once.	0	0	1	1
4.	If there is an interruption, I can switch back to what I was doing very quickly.	0	0	1	1
5.	I find it easy to "read between the lines" when someone is talking to me.	0	0	1	1
6.	I know how to tell if someone listening to me is getting bored.	0	0	1	1
7.	When I'm reading a story, I find it difficult to work out the characters' intentions.	1	1	0	0
8.	I like to collect information about categories of things (e.g., types of cars, types of birds, types of trains, types of plants, etc.).	1	1	0	0
9.	I find it easy to work out what someone is thinking or feeling just by looking at their face.	0	0	1	1
10.	I find it difficult to work out people's intentions.	1	1	0	0

Note that the AQ is not diagnostic, although it is recommended by the guidelines of the UK's National Institute for Health and Care Excellence (NICE) as a screening instrument for autism.

The average score for females is 3.16 (the range is 1 to 5), and the average score for males is 3.56 (the range is 1 to 6). If you score 0 to 4, you are low in autistic traits, and if you score 6 to 10, you are high in autistic traits. If you scored above 6, you are struggling, your struggles predated taking the test, and you think you might benefit from the support that an autism diagnosis might bring, then ask your family physician for a referral to a specialist clinic for a diagnosis.[1]

Acknowledgments

My wife Bridget Lindley died unexpectedly during the writing of this book. She gave our family so much love over decades, and our children, Sam, Kate, and Robin, were beyond amazing as we supported each other through this huge loss. Their partners, Alice Seabright and Alex Rodin, and my siblings, Dan, Ash, and Liz, were also a tremendous source of support. My friend Lucy Richer, who loves stories, particularly those giving a voice to the voiceless, encouraged me back into writing. When I did, my editors Josephine Greywoode, Helen Conford, and Thomas Kelleher; my literary consultant Robin Dennis; and my agents Katinka Matson and John Brockman, gave me invaluable feedback.

Many talented past and present graduate students, postdocs, and colleagues helped me with the research behind this book: Dwaipayan (Deep) Adhya, Carrie Allison, Chris Ashwin, Topun Austin, Bonnie Auyeung, Ezra Aydin, Richard Bethlehem, Jaclyn Billington, Thomas Bourgeron, Ed Bullmore, Bhismadev Chakrabarti, Tony Charman, Emma Chapman, Adriana Cherskov, Lindsay Chura, Jamie Craig, Dorothea (Dori) Floris, Jan Freyberg, Lidia Gabis, Gina Gomez de la Cuesta, Davíð Guðjónsson, Ofer Golan, David Greenberg, Sarah Griffiths, Sarah Hampton, Rosie Holt, Nazia Jassim, Therese Jolliffe, Rebecca Knickmeyer, Mariann Kovacs, Meng-Chuan Lai,

Johnny Lawson, Mike Lombardo, Svetlana Lutchmaya, Aicha Massrali, Michelle O'Riordan, Owen Parsons, Arko Paul, Alexa Pohl, Wendy Phillips, Tanya Procyshyn, Howard Ring, Caroline Robertson, Janine Robinson, Amber Ruigrok, Liliana Ruta, Emily Ruzich, Fiona Scott, Paula Smith, John Suckling, Sophia Sun, Teresa Tavassoli, Alex Tsompanides, Florina Uzefofsky, Varun Warrier, Elizabeth Weir, and Sally Wheelwright. Our amazing admin team, Emma Baker, Anna Crofts, Joanna Davis, Becky Kenny, and Aubree Wisley, kept my lab running smoothly to give me space to write. To all of them, I say a warm thank-you.

In particular, I am indebted to Lucy Richer, Carrie Allison, Jon Drori, Mike Lombardo, Varun Warrier, Lauri Love, Daniel Tammet, David Greenberg, Imre Leader, Tanya Procyshyn, Nicholas Conard, Shankar Balasubramanian, and Adam Ockelford, who gave me really helpful feedback on the draft of this book through their careful reading and engaging with the ideas.

The first draft of this book was written during a writing retreat with my son Sam, on Lac Masson in the beautiful Laurentian Mountains in Quebec, overlooking the lake. The final draft was written during a second writing retreat with my son Robin and daughter Kate, in Treyarnon Bay in Cornwall, overlooking the wild waves of the Atlantic and the equally beautiful Cornish cliffs. And I thank my parents, Judy and Vivian, who encouraged a love of learning in their children.

As I finish this book, on March 24, 2020, when governments around the world have declared a state of emergency in the war against the biggest viral pandemic for a century, we humans remain connected even in a state of physical isolation by using video phone calls that hyper-systemizers invented. Our thanks should go to people like Janus Fris, the Danish co-inventor of Skype, who had no formal education and dropped out of high school. And night and day, in molecular biology laboratories, hyper-systemizers are working tirelessly to invent a new vaccine that will outsmart the invisible killer that is COVID-19, the new coronavirus. And as we look beyond COVID-19 at the other major challenge our planet faces—climate destruction—we look to hyper-systemizers to invent new solutions.

The great physicist Freeman Dyson wrote a book called *Maker of Patterns* to describe his life exploring patterns in mathematics and physics. This book joins the dots between pattern-seeking in our world's greatest scientists and inventors, and pattern-seeking in autistic people. Autistic people, even if not formally diagnosed, often hide in the shadows, avoiding the limelight, and systemize to such an extreme that they are more likely to invent something, though they may not even put their name to their invention. They just systemize for the pure pleasure of systemizing. Systemizing is wired into our brains by evolution, and it is wired into the brains of autistic people to such a high degree that they can't help but do it all day long.

Psychologist Steven Pinker writes in *The Language Instinct* about how language is an instinct for humans just as much as web-spinning is an instinct for spiders:

> "Web-spinning was not invented by some unsung spider genius and does not depend on having had the right education or on having an aptitude for architecture or the construction trades. Rather, spiders spin webs because they have spider brains, which give them the urge to spin and the competence to succeed."

Autistic people systemize not because they are driven by ego or a quest for fame or fortune (some may be, but in my experience most are not), and not because a schoolteacher has told them to systemize. Rather, they systemize because this is what evolution has designed their brains to do. They systemize for the sheer pleasure of detecting *if-and-then* patterns, patterns they often see almost effortlessly while the rest of us may labor to spot it at all.

To every autistic person, and to your families, I extend a warm thank-you. Science has confirmed my everyday experience of meeting you: that even if you struggle with cognitive empathy, you are more moral than others, because you combine affective empathy with a strong love of logic and an overriding belief in fairness and justice. You, like countless other autistic people who walked the earth before you, including many who were never formally diagnosed, understand how things work, having identified patterns and introduced one tiny modification to them at a time. In so doing, you have invented.

Now, thanks to the genes for systemizing passed on to you by your ancestors of 70,000 to 100,000 years ago, I can send a text message to you in a split-second, anywhere on the planet, carried invisibly on the waves.

Notes and Further Reading

CHAPTER 1: BORN PATTERN SEEKERS

1. I sat in on the assessment of Jonah. Naturally, the details have been anonymized to protect confidentiality.

2. For an online resource on vocabulary development, see the data on vocabulary development from thousands of kids compiled by Stanford scientists and made available for other researchers to use at wordbank.stanford.edu /analyses?name=vocab_norms.

3. See R. Reader et al. (2014), "Genome wide studies of specific language impairment," *Current Behavioural Neuroscience Reports* 1(4), 242–250.

4. On the link between late-talking children and parents who are engineers, see T. Sowell (1998), *Late-talking children* (New York: Basic Books).

5. The phrase "exact mind" comes from S. Baron-Cohen and S. Wheelwright (2004), *An exact mind: An artist with Asperger syndrome*, artwork by Peter Myers (London: Jessica Kingsley Ltd.).

6. It turned out that Jonah's ability to classify leaves was scientifically accurate. See M. Hickey and C. Clive (1997), *Common families of flowering plants* (Cambridge: Cambridge University Press).

7. On repetitive behavior, see I. Carcani-Rathwell et al. (2006), "Repetitive and stereotyped behaviors in pervasive developmental disorders," *Journal of Child Psychology and Psychiatry* 47(6), 573–581. A useful review of the term "RRBI" can also be found at Interactive Autism Network, "Autism: Restricted and repetitive behaviors," April 2, 2007, updated November 7, 2013, iancommunity.org/cs /autism/restricted_repetitive_behaviors.

8. Until the early part of the twenty-first century, repetitive behavior in autism was seen as negative and unwanted, and teachers and clinicians advised parents to discourage a child's repetitive behavior. I challenged this view by reinterpreting repetitive behavior as "systemizing," an intelligent and—as we

shall see—uniquely human behavior reflecting a different learning style. See S. Baron-Cohen (2002), "The extreme male brain theory of autism," *Trends in Cognitive Sciences* 6, 248–254.

9. These details about Edison are taken from Gerald Beals, "The biography of Thomas Edison," June 1999, www.thomasedison.com/biography.html.

10. According to dictionary definitions, "invention" involves coming up with something new: a new method, a new idea, a new device, a new discovery, a new process. Indeed, the root of the word is "-vent," from the Latin *venire*, which means "to come." Some definitions contrast the term "invention" with the related term "innovation," and one meaning of innovation is an invention that is taken up widely across the population. Others define innovation as building on an existing invention. For me, whether an invention takes off across the population may depend on social and economic factors, such as whether the inventor has access to funding or a good marketing team. The much more basic question is to ask how it is that we humans are capable of invention, however mundane or important, and irrespective of whether our inventions succeed.

11. The phrase "landmark moment" is not to be taken literally, of course. The 30,000-year window I've estimated for the emergence of the Systemizing Mechanism (between 70,000 to 100,000 years ago) allows for both abrupt and incremental models of neural change.

CHAPTER 2: THE SYSTEMIZING MECHANISM

1. The term "Systemizing Mechanism" was first introduced in S. Baron-Cohen (2006), "Two new theories of autism: Hyper-systemizing and assortative mating," *Archives of Diseases in Childhood* 91, 2–5, and explored further in S. Baron-Cohen (2006), "The hyper-systemizing, assortative mating theory of autism," *Progress in Neuropsychopharmacology and Biological Psychiatry* 30, 865–872.

2. The term "systemizing" was first introduced in S. Baron-Cohen (2002), "The extreme male brain theory of autism," *Trends in Cognitive Sciences* 6, 248–254, and operationalized in S. Baron-Cohen et al. (2003), "The Systemizing Quotient: An investigation of adults with Asperger syndrome or high-functioning autism, and normal sex differences," *Philosophical Transactions of the Royal Society: Series B* 358, 361–374.

3. I've deliberately not included a "who" question, because that relates to people rather than objects and inanimate events, and we typically ask questions about people's behavior through a different mechanism, the Empathy Circuit. This is discussed later in chapter 2.

4. See A. Gopnik (2002), "Why do we ask questions?," Edge, www.edge.org /response-detail/11928; and D. Premack and A. J. Premack (1983), *The mind of an ape* (New York and London: W. W. Norton & Co.).

5. See M. Chouinard et al. (2007), "Children's questions: A mechanism for cognitive development," *Monographs of the Society for Research in Child Development* 72(1), 1–112.

6. See R. Proctor (2012), "The history of the discovery of the cigarette–lung cancer link: Evidentiary traditions, corporate denial, global toll," *Tobacco Control* 21, 87–91; and R. Doll and A. Hill (1954), "The mortality of doctors in relation to their smoking habits," *British Medical Journal* 1, 1451–1455.

7. See T. Maciel (2015), "The physics of sailing: How does a sailboat move upwind?" Physics Central, May 12, 2015, physicsbuzz.physicscentral.com/2015/05/the-physics-of-sailing-how-does.html.

8. Note that some systems are designed entirely independently to produce the same output, that is, to have the same function. For example, in some cultures and at some points in history, children are taught to do long multiplication in math using the Mayan system, while in other cultures or at other points in history, children are taught to do this using the lattice system, and both ways may be equally efficient. For the lattice system, see Len Goodman, "Lattice method," MathWorld, mathworld.wolfram.com/LatticeMethod.html. For the Mayan system, see "Mayan mathematics," The Story of Mathematics, www.storyofmathematics.com/mayan.html.

9. See R. Castleden (1993), *The making of Stonehenge* (New York: Routledge).

10. See D. Anthony (2007), *The horse, the wheel, and language* (Princeton, NJ: Princeton University Press).

11. This example can be thought of literally as "horse power," although not in the way that James Watt, who invented steam engines, used the term. Watt calculated a mathematical way to equate horses to engine power. He measured the capability of a strong horse to pull a load. The conventional definition of "horsepower" is the power needed to pull 330 pounds 100 feet in one minute, or 33 pounds 1,000 feet in one minute, or 1,000 pounds 33 feet in one minute. In other words, one horsepower equals 33,000 foot-pounds in one minute.

12. *And* is often a placeholder for the ultimate cause of a change, when it is finally isolated and understood. So, to answer the "why" question "Why did the candle go out?," we hypothesize and confirm the pattern "*if* the candle is lit, *and* there is sudden strong wind, *then* the candle goes out." The *and* term (the sudden strong wind) is a proxy for the cause of the candle going out. It means that, so far, this is as close as this human observer, or their brain, has been able to get to identifying the actual causal operation. Scientists are famously extremely careful not to credit something as being causal until it is proven beyond reasonable doubt. The Systemizing Mechanism was the result not of professional scientific training but of evolution; as such, it provides a neural mechanism that can try to identify the causes of things, to a best approximation. See P. Lipton (1991), *Inference to the best explanation* (London: Routledge). I am indebted to my friend Peter Lipton, whom I first encountered when he was a PhD student in New College Oxford and I was an undergraduate in the same college, looking for tutorials in philosophy of science. He invited me to his college room and asked me to define the word causality. Like any excellent teacher, he sat silently while I thought about it, also silently, for what seemed like five minutes. Some teachers rush in to help their student when they don't seem to be able to answer,

but Peter was content just to let the sun go down and wait, so that I could collect my own thoughts. I was delighted when, fifteen years later, we met up again, he a lecturer of philosophy in Cambridge, and I a lecturer there in psychology. He had recently published his book, and we went for lunch in Kings College Cambridge and discussed the concept of inference, which went back at least as far as Aristotle in 300 BC. On my account, systemizing involves "inference, to the best explanation" but dates back to between 70,000 and 100,000 years ago. Tragically, Peter died way too young, age fifty-three.

13. The uniqueness of systemizing to humans is explored more fully in chapter 5, but see D. Povinelli and S. Dunphy-Lelii (2001), "Do chimpanzees seek explanations? Preliminary comparative investigations," *Canadian Journal of Experimental Psychology* 55(2), 187–195.

14. See C. Gibbs-Smith (1962), *Sir George Cayley's aeronautics, 1796–1835* (London: HM Stationery Office).

15. Note that the *and* is not necessarily a causal operation here, but (as explained in note 12 above) more likely a proxy for a causal operation. The Systemizing Mechanism simply seeks *if-and-then* patterns and identifies how reliable they are and treats something as causal even if it is not the ultimate cause.

16. See J. Sumner (2000), *The natural history of medicinal plants* (Portland, OR: Timber Press). We now understand that willow bark is effective in relieving headache pain because it contains salicylic acid, which we now call aspirin.

17. Dogs eat grass to help them vomit if they have a parasite, and the monarch butterfly lays its eggs on milkweed, which has anti-parasite effects. But they likely do this without knowing why they are doing it, either because they have formed a learned association or for genetic reasons. This is not the same as systematically experimenting with different foods as medicines. See J. Shurkin (2014), "Animals that self-medicate," *Proceedings of the National Academy of Sciences* 111(49), 17339–17341.

18. On how "primitive" life could have been quite comfortable 70,000 to 100,000 years ago, see the popular "What is primitive technology?," available on YouTube and described by G. Pierpoint, "What is 'primitive technology' and why do we love it?," BBC News, August 27, 2018, www.bbc.co.uk/news/blogs-trending -45118653; for the longer version, see "Primitive Technology," www.youtube .com/channel/UCAL3JXZSzSm8AlZyD3nQdBA/featured.

19. See H. Chisholm, ed. (1911), "Boole, George," in *Encyclopedia Britannica*, 11th ed. (Cambridge: Cambridge University Press). One of his important works was G. Boole (1854), *An investigation of the laws of thought* (London: Walton & Maberly). When scientists want to look for patterns in big data, or when they program computers to do so, the algorithms they use tend to follow the same *if-and-then* Boolean systemizing format. An example is when a scientist is trying to extract rules from tens of thousands of animals as to whether to classify them as mammals or reptiles. The computer will crawl through the data to find the best algorithm to separate exemplars into these two (or more) categories, such as *if* it gives birth, *and* it is warm-blooded, *then* = mammal; but *if* it does not give birth, *and* it is cold-blooded, *then* = reptile. This example is taken from a course

handout on classifiers, "Data mining: Rule-based classifiers," staffwww.itn.liu
.se/~aidvi/courses/06/dm/lectures/lec4.pdf.

20. On Mary Boole, see P. Nahin (2012), *The logician and the engineer: How George
Boole and Claude Shannon created the information age* (Princeton, NJ: Princeton University Press), 28. And see her own book, M. E. Boole (1909), *Philosophy and fun
of algebra* (London). On George Boole's death, see Tommy Barker, "Have a look
inside the home of UCC maths professor George Boole," *Irish Examiner*, June
13, 2015; see also S. Burris, "George Boole," in *The Stanford encyclopedia of philosophy*, April 21, 2010, updated April 18, 2018, plato.stanford.edu/entries/boole/;
and J. J. O'Connor and E. F. Robertson, "George Boole," www-groups.dcs.st
-and.ac.uk/history/Biographies/Boole.html.

21. See A. Gopnik et al. (2001), "Causal learning mechanisms in very young
children: Two, three and four year olds infer causal relations from patterns of
variation and covariation," *Developmental Psychology* 37, 620–629; see also D. Sobel
et al. (2004), "Children's causal inferences from indirect evidence: Backwards
blocking and Bayesian reasoning in preschoolers," *Cognitive Science* 28, 303–333;
and A. Gopnik et al. (1999), *How babies think* (London: Weidenfeld and Nicolson).

22. See D. A. Lagnado et al. (2007), "Beyond covariation: Cues to causal structure," in A. Gopnik and L. Schulz, eds., *Causal learning: Psychology, philosophy,
and computation* (Oxford: Oxford University Press); L. Schulz et al. (2007), "Preschool children learn about causal structure from conditional interventions,"
Developmental Science 10, 322–332; and C. Lucas et al. (2014), "When children are
better (or at least more open-minded) learners than adults: Developmental differences in learning the forms of causal relationships," *Cognition* 131, 284–299.

23. See American Botanical Society, "The mysterious Venus fly trap," www
.botany.org/bsa/misc/carn.html.

24. Systemizing only explains curiosity about causal or systemizable events.
Curiosity about social events, in my view, is the result of a different uniquely
human cognitive mechanism, the Empathy Circuit.

25. See R. Smith (2013), "World's oldest calendar discovered in the UK,"
National Geographic, July 16.

26. Mesopotamia turns out to have been a remarkable location for early systemizing. In Mesopotamia they also developed the earliest writing system, called
cuneiform, and developed a sexagesimel (base 60) number system to help record
very big or very small numbers; just as we divide a circle into 360 degrees today,
subdivided into 60 minutes, so did they. See A. Asger (1991), "The culture of
Babylonia: Babylonian mathematics, astrology, and astronomy," in J. Boardman
et al., eds., *The Assyrian and Babylonian empires and other states of the Near East, from
the eighth to the sixth centuries BC* (Cambridge: Cambridge University Press). Astonishing inventions such as the wheel and agriculture also all happened in what
is called the Fertile Crescent, which spans Mesopotamia, Palestine, and Egypt.

27. On Halley's Comet, see G. W. Kronk (1999), *Cometography*, vol. 1,
Ancient–1799 (Cambridge: Cambridge University Press). The earliest observations of the planet Venus were recorded around 1700 BC. These observations
document the rise time of Venus over a twenty-one-year period. The ancient

Greeks built on this, systemizing the nighttime sky. Aristarchus of Samos estimated the size and distance of the moon and the sun in 3 BC and proposed the heliocentric model of the solar system, although Copernicus got the credit for it in the sixteenth century. See F. Espenak (2004), "Transits of Venus, six millennium catalog: 2000 BCE to 4000 CE," NASA, February 11.

28. On the *Zhou Shu*, see G. Chambers (1899), *The story of eclipses* (London: George Newnes Ltd.). The book is part of a compendium known as Zhou Shu (or "Book of Zhou"). See Edward L. Shaughnessy (1999), "Western Zhou history," in M. Loewe and E. L. Shaughnessy, eds., *The Cambridge history of ancient China: From the origins of civilization to 221 B.C.* (Cambridge: Cambridge University Press).

29. On lunar eclipses, see E. Livni (2018), "The terrifying history of lunar eclipses," *Quartz*, July 26.

30. Linnaeus wrote in his journal that he "read day and night, knowing like the back of my hand, Arvidh Månsson's Rydaholm *Book of herbs*, Tillandz's *Flora Åboensis*, Palmberg's Serta Florea Suecana, Bromelii Chloros Gothica and Rudbeckii Hortus Upsaliensis." See "Linnaeus, Carl," All About Heaven, allaboutheaven .org/sources/linnaeus/190.

31. If we take an example: Since Linnaeus, we now classify birds in a class of animals called Aves. The class of Aves itself is split up into 23 orders, the biggest of these being passerines. Orders are then divided into families (like the Apodidae, or typical swifts), and each family is then subdivided into genuses—there are 2,057 genuses in the class of Aves. Genuses are then subdivided into species—there are 9,702 species of birds. The final layer of classification is the subspecies, for a species that differs slightly in one geographical area. See "Bird classifications," Birds.com, www.birds.com/species/classifications/. Notice that when it comes to inventing a classification system, the *and* in the *if-and-then* pattern isn't necessarily causal, although the *and* is serving a similar role in adding a feature to change one bird into another.

32. A variety of subtly different terms for bird-watchers are used. The following definitions appeared in the "Birding glossary" (1969) of *Birding* magazine (vol. 1, no. 2): A "birder" is a person who seriously pursues the hobby of birding. "Birding" is a hobby in which individuals enjoy the challenge of bird study and bird listing. A "bird-watcher" is an ambiguous term used to describe the person who watches birds for any reason at all, not to be confused with the serious birder. "Twitching" is a British term used to mean "the pursuit of a previously located rare bird." The term "twitcher" refers to those who travel long distances to see a rare bird, which is then "ticked," or counted, on a list. See P. Dunne (2003), *Pete Dunne on bird watching* (Boston: Houghton Mifflin).

33. On systemizing apples, see University of Illinois Extension, "Apples and more," extension.illinois.edu/apples/facts.cfm. Note that there are many ways to systemize an apple, one being by taste. See "The spectrum of apple flavors," Blame It on the Voices, July 10, 2010, www.blameitonthevoices.com/2010/07 /know-your-apples-spectrum-of-apple.html.

34. In my view, the Systemizing Mechanism is different to what is called "statistical learning." Statistical learning is about learning regularities and typically occurs in the absence of an obvious external reward, unlike associative learning. For example, Saffron and her colleagues reported that human infants can identify statistical regularities in speechlike sounds and concluded that infants are "natural born statisticians." Importantly, infants can spot regularities not only in speechlike sounds but in simple musical tones and in visual shapes. The human infant's ability to do statistical learning is impressive and likely was an evolutionary precursor to systemizing, but is not the same as systemizing. This is because it is possible to do statistical learning without needing *if-and-then* reasoning. Thus, monkeys and rats can do statistical learning, yet as we will see in chapter 6, they cannot invent generatively. In statistical learning, all you have to do is remember that A is paired with B, and how often, without needing to understand *if-and-then* patterns. Statistical learning is also sometimes referred to as "probabilistic learning." See R. Aslin (2017), "Statistical learning: A powerful mechanism that operates by mere exposure," *WIRES (Wiley Interdisciplinary Reviews): Cognition and Science* 8(1–2), 1–7; J. Saffron et al. (1996), "Statistical learning by 8 month old infants," *Science* 274, 1926–1928; J. Saffron et al. (1999), "Statistical learning of tone sequences by human infants and adults," *Cognition* 70, 27–52; N. Kirkham et al. (2002), "Visual statistical learning in infancy: Evidence for a domain general learning mechanism," *Cognition* 83, B35–B42; M. Hauser et al. (2001), "Segmentation of the speech stream in a non-human primate: Statistical learning in cotton-top tamarins," *Cognition* 78, B53–B64; and C. Santolin and J. Saffron (2018), "Constraints on statistical learning across species," *Trends in Cognitive Science* 22(1), 52–63.

35. On associative learning, see B. F. Skinner (1938), *Behavior of organisms* (New York: Appleton-Century-Crofts).

36. An example of a mental spreadsheet is how our brain systemizes cars using make and registration plate to recognize that the car that just drove by is Sarah's car: *if* the number plate is AIE7JY, *and* it's a Renault Laguna, *and* it's red, *then* it's Sarah's car. Similarly, the brain systemizes by making maps of *what* happened, *when* it happened, and *where* it happened in space and time, the bedrock of our physical universe. See G. Buzsáki and R. Llinás (2017), "Space and time in the brain," *Science* 358(6362), 482–485.

37. Gardeners know that to make the soil more acidic and change the rhododendron's flower color, they should mix in one or two inches of sphagnum peat moss. See K. Adams, "What can you use to change the color of a rhododendron flower?," SFGate, homeguides.sfgate.com/can-use-change-color-rhododendron-flower-68727.html.

38. Here's the algorithm for systemizing the menstrual cycle: "*if* it's four or five days before I'm due to ovulate, *and* there's an increase in my body temperature, *then* I could get pregnant." In this rule, you can add more *and* clauses, such as "*and* I detect blood spots, *and* I detect increased cervical fluid." See D. Dunnington, "The menstrual cycle and sleep," SleepHub, August 17, 2015, sleephub

.com.au/menstrual-cycle-and-sleep/. Here's the algorithm for systemizing rocks: *if* it's an old rock, *and* it has been squeezed but not melted, *then* it's metamorphic. See "Identifying rocks," Science 6 at FMS, June 7, 2012, fitz6.wordpress.com/2012/06/07/identifying-rocks/.

39. Some biographical research suggests that Isaac Newton and other outstanding physicists and scientists may have been autistic—long before the diagnosis was available. Retrospective diagnosis when the person is no longer alive is replete with difficulties, as the evidence may be fragmentary and the person or their family cannot give complete accounts. Nevertheless, see I. James (2003), "Singular scientists," *Journal of the Royal Society of Medicine* 96(1), 36–39.

40. The astronomer who systemized tidal patterns 3,000 years ago was Aristarchus of Samos. See T. Heath (1913), *Aristarchus of Samos, the ancient Copernicus* (London: Oxford University Press). The form of the algorithm for tidal patterns would be, for example, *if* on Saturday at 10:00 a.m. the tide height is low, *and* the time changes to Saturday at 2:00 p.m., *then* the tide height will rise. Here's an example of systemizing the shape of waves: *if* you take the length of the wave divided by its width, *and* it's less than three, *then* it's an "almond"-tube-shaped wave. See "Surfing," Wikipedia, en.wikipedia.org/wiki/Surfing#/media/File:Wave-shape-intensity.svg.

41. On systemizing the skateboard, see "Skateboard trick list," www.skateboardhere.com/skateboard-trick-list.html/. The gender split among skateboarders is currently about 80 percent male, 20 percent female; see "Who are skateboarders?," Public Skateboard Development Guide, publicskateparkguide.org/vision/who-are-skateboarders/.

42. Fleming quoted in K. Haven (1994), *Marvels of science: 50 fascinating 5-minute reads* (Littleton, CO: Libraries Unlimited); see also L. Colebrook (1956), "Alexander Fleming 1881–1955," *Biographical Memoirs of Fellows of the Royal Society* 2, 117–126; and R. Cruickshank (1955), "Sir Alexander Fleming, FRS," *Nature* 175(4459), 663.

43. On the intraparietal sulcus, see G. A. Orban et al. (2006), "Mapping the parietal cortex of human and non-human primates," *Neuropsychologia* 44(13), 2647–2667; D. Stout and T. Chaminade (2007), "The evolutionary neuroscience of tool making," *Neuropsychologia* 45(5), 1091–1100; D. Stout et al. (2008), "Neural correlates of early Stone Age tool-making: Technology, language, and cognition in human evolution," *Philosophical Transactions of the Royal Society: Series B* 363(1499), 1939–1949; and K. Kucian et al. (2006), "Impaired neural networks for approximate calculation in dyscalculic children: A functional MRI study," *Behavior and Brain Function* 2, 31.

44. On the brain basis of systemizing, see S. Baron-Cohen and M. V. Lombardo (2017), "Autism and talent: The cognitive and neural basis of systemizing," *Translational Research* 19(4), 345–353.

45. On systemizing people's behavior, see S. Baron-Cohen (2011), *Zero degrees of empathy* (London: Penguin UK), published in the United States as *The science of evil* (2012) (New York: Basic Books).

46. Lawyers experiment to change society. An example might be, *if* a seventeen-year-old is refusing a lifesaving blood transfusion, *and* the Children's Act says that the court must protect a person under eighteen years old, *then* the court can order the hospital to give the seventeen-year-old the blood transfusion, even if against his wishes. This example is the basis of the plot in the novel *The Children Act* (2014) by Ian McEwan.

47. It has been noted that some autistic females, or some females with undiagnosed autism, may become "obsessed" with film or novels, and have an excellent understanding of the drama and even harness this in their work to become experts, even though they may struggle to have an ordinary conversation in real life and may avoid social situations that they find too challenging. This may be one way in which they "camouflage" their autism, effectively learning theory of mind from the "static" or repeatable world of books or movies even if implementing it in real time in the social world remains challenging. See L. Hull et al. (2017), "'Putting on my best normal': Social camouflaging in adults with autism spectrum conditions," *Journal of Autism and Developmental Disorders* 47, 2519–2534; M.-C. Lai et al. (2016), "Quantifying and exploring camouflaging in men and women with autism," *Autism* 21, 690–702; L. Hull (2018), "Development and validation of the camouflaging autistic traits questionnaire (CAT-Q)," *Journal of Autism and Developmental Disorders* 1, 5; M.-C. Lai et al. (2018), "Neural self-representation in autistic women and association with 'compensatory camouflaging,'" *Autism* 23(5), 1210–1223; L. Hull et al. (2019), "Gender differences in self-reported camouflaging in autistic and non-autistic adults," *Autism* 24, 352–363; L. Livingstone et al. (2019), "Good social skills despite poor theory of mind: Exploring compensation in autism spectrum disorder," *Journal of Child Psychology and Psychiatry* 60(1), 102–110.

48. On other animals' ability to mindread, see C. Heyes (2015), "Animal mindreading: What's the problem?," *Psychonomic Bulletin and Review* 22(2), 313–327; D. Premack and G. Woodruff (1978), "Does the chimpanzee have a theory of mind?," *Behavioral and Brain Sciences* 1(4), 515–526; J. Call and M. Tomasello (2008), "Does the chimpanzee have a theory of mind? 30 years later," *Trends in Cognitive Sciences* 12, 187–192. Call and Tomasello argue that while chimpanzees may understand volitional mental states like having a desire or a goal, there is still no robust evidence that they understand epistemic mental states like belief, especially false belief.

49. See A. Whiten and R. Byrne (2010), "Tactical deception in primates," *Behavioural and Brain Sciences* 11(2), 233–244. In this landmark discussion of the topic, a number of scientists argued that learning a paired association between A and B could give rise to what seems like deception in another animal. An example of such an A-and-B rule might be "another animal is present" paired with "eat your food behind a rock." A rule like this may have survival value, as an animal would be less likely to lose its food if it followed it. But this is very different to deception on the order of "I want the other animal to *believe* I don't have any food."

50. See J. Hoffecker and I. Hoffecker (2017), "Technological complexity and the dispersal of modern humans," *Evolutionary Anthropology* 26, 285–299. This article discusses how *Homo sapiens* laid traps and snares and used darts as silent weapons, but other hominids did not. This shows the need for both a Systemizing Mechanism and an Empathy Circuit, working in tandem. See chapter 5, note 16, for the controversy over whether Neanderthals were capable of this.

51. On teaching in other species, see K. N. Laland (2017), *Darwin's unfinished symphony* (Princeton, NJ: Princeton University Press); and A. Thornton and K. McAuliffe (2006), "Teaching in wild meerkats," *Science* 313, 227–229.

52. On theory of mind in language, see H. P. Grice (1989), *Studies in the way of words* (Cambridge, MA: Harvard University Press); and J. L. Austin (1962), *How to do things with words: The William James Lectures delivered at Harvard University in 1955*, ed. J. O. Urmson and M. Sbisà (Oxford: Clarendon Press).

53. See C. Colonnesi et al. (2010), "The relation between pointing and language development: A meta-analysis," *Developmental Review* 30(4), 352–366; M. Tomasello (2006), "Why don't apes point?," in *Roots of human sociality: Culture, cognition, and interaction*, ed. N. Enfield and S. Levinson (Oxford and New York: Berg); and A. Smet and R. Byrne (2013), "African elephants can use human pointing cues to find hidden food," *Current Biology* 23(20), 2033–2037. Other animals may understand some forms of pointing as a directional cue, but there is limited evidence that they understand it as an intention to refer. See A. Miklosi and K. Soproni (2006), "A comparative analysis of animals' understanding of the human pointing gesture," *Animal Cognition* 2, 81–93.

54. On empathy in other species, see F. De Waal (2005), "The empathic ape," *New Scientist*, October 8; and I. Ben-Ami Bartal et al. (2011), "Empathy and pro-social behavior in rats," *Science* 334, 1427.

55. On the false-belief test, see H. Wimmer and J. Perner (1983), "Beliefs about beliefs: Representation and constraining function of wrong beliefs in young children's understanding of deception," *Cognition* 13, 103–128; and S. Baron-Cohen et al. (1985), "Does the autistic child have a 'theory of mind'?," *Cognition* 21, 37–46. On the brain basis of mindreading, see C. Wiesemann et al. (2020), "Two systems for thinking about other thoughts in the developing brain," *Proceedings of the National Academy of Sciences*, March 9; and S. Baron-Cohen (2011), *Zero degrees of empathy*.

56. See C. Krupenye et al. (2016), "Great apes anticipate that other individuals will act according to false beliefs," *Science* 354(6308), 110–114.

57. See M. Balter (2013), "Are crows mindreaders, or just stressed out?," *Science*, January 10.

58. See Y. Tomonaga et al. (2010), "Bottlenose dolphins' (*Tursiops truncatus*) theory of mind as demonstrated by responses to their trainers' attentional states," *International Journal of Comparative Psychology* 23, 386–400.

CHAPTER 3: FIVE TYPES OF BRAIN

1. See D. Greenberg et al. (2018), "Testing the Empathizing-Systemizing (E-S) theory of sex differences and the Extreme Male Brain (EMB) theory of autism in more than half a million people," *Proceedings of the National Academy of Sciences* 115(48), 12152–12157. In the UK Brain Types Study, we used brief (ten-item) versions of the two questionnaires, the Empathy Quotient (EQ-10) and the revised Systemizing Quotient (SQ-R-10). The SQ-R focuses on systems found not just in STEM fields but in everyday life (for example, the weather, maps, mountains, and furniture).

2. Bus and train timetables are a clear example of information that follows the *if-and-then* structure. For example, *if* the bus left York at 7:00 a.m., *and* it is a weekday, *then* it will reach Whitby at 8:10 a.m. Note that such timetables were being produced long before the invention of the digital spreadsheet. An example algorithm for systemizing cooking might be: "*if* I take the dough, *and* add more yeast, *then* the bread I bake will be taller." The invention of cooking was transformative for humans and would have emerged as a result of the Systemizing Mechanism. An example algorithm for systemizing bicycle mechanics might be: "*if* I have sit-up handlebars on my bike, *and* I change these for drop handlebars, *then* the bike will go faster." An example algorithm for systemizing public health might be: *if* people are in contact with fecal bacteria, *and* they use soap to wash their hands, *then* they are less likely to develop diarrhea. In this example, tweaking the operation (the *and* step) repeatedly led to a huge difference in health outcomes. It is thought that handwashing reduces diarrheal illness by around 30 percent, and by 43 to 47 percent if soap is used. This led public health scientists to conclude that handwashing with soap may be "the single most cost-effective way of reducing the global burden of infectious disease." See V. Curtis et al. (2011), "Hygiene: New hopes, new horizons," *Lancet Infectious Diseases* 11(4), 312–321.

3. If a trait shows a bell curve (or is normally distributed), this doesn't prove it is polygenic, since normal distributions can emerge for nongenetic reasons. To prove a trait is genetic, one first needs evidence that it is heritable, which twin studies can provide. But if a trait is polygenic, a normal distribution will emerge if each of the hundreds or thousands of common variants are having a small effect, and where the chance of having any one common genetic variant is binary (like tossing a coin). For an explanation, see K. Oldenbroek and L. van der Waaij (2014), *Textbook animal breeding: Animal breeding and genetics for BSc students* (Wageningen, Netherlands: Centre for Genetic Resources and Animal Breeding and Genomics Group, Wageningen University and Research Centre), chapter 5.4, "Polygenic genetic variation," wiki.groenkennisnet.nl /display/TAB/Chapter+5.4+Polygenic+genetic+variation.

4. On the five brain types, see A. Wakabayashi et al. (2007), "Empathizing and systemizing in adults with and without autism spectrum conditions: Cross-cultural stability," *Journal of Autism and Developmental Disorders* 37, 1823–1832; and Y. Groen et al. (2015), "The Empathy and Systemizing Quotient: The

psychometric properties of the Dutch version and a review of the cross-cultural stability," *Journal of Autism and Developmental Disorders* 45(9), 2848–2864.

5. See D. Treffert (2009), "The savant syndrome: An extraordinary condition: A synopsis: Past, present, future," *Philosophical Transactions of the Royal Society of London: Series B, Biological Sciences* 364(1522), 1351–1357; K. Hyltenstam (2016), *Advanced proficiency and exceptional ability in second languages* (Berlin: Walter de Gruyter GmbH); D. Kennedy and L. Squire (2007), "An analysis of calendar performance in two autistic calendar savants," *Learning and Memory* 14(8), 533–538.

6. A zero-sum game would be expected if empathy and systemizing competed for some common biological resource. See N. Goldenfeld et al. (2005), "Empathizing and systemizing in males, females, and autism," *Clinical Neuropsychiatry* 2, 338–345; and N. Goldenfeld et al. (2007), "Empathizing and systemizing in males, females, and autism: A test of the neural competition theory," in T. Farrow, ed., *Empathy and mental illness* (Cambridge: Cambridge University Press).

7. On autism and systemizing mechanical systems, see S. Baron-Cohen et al. (2001), "Studies of theory of mind: Are intuitive physics and intuitive psychology independent?," *Journal of Developmental and Learning Disorders* 5, 47–78.

8. See X. Wei et al. (2013), "Science, Technlogy, Engineering, and Mathematics (STEM) participation among college students with an autism spectrum disorder," *Journal of Autism and Developmental Disorders* 43, 1539–1546.

9. On autism and pattern recognition, see L. Mottron et al. (2009), "Enhanced perception in savant syndrome: Patterns, structure, and creativity," *Philosophical Transactions of the Royal Society of London: Series B, Biological Sciences* 364(1522), 1385–1391; I. Soulières et al. (2009), "Enhanced visual processing contributes to matrix reasoning in autism," *Human Brain Mapping* 30(12), 4082–4107; U. Frith (1972), "Cognitive mechanisms in autism: Experiments with colour and tone sequence production," *Journal of Autism and Childhood Schizophrenia* 2, 160–173.

10. See S. Baron-Cohen et al. (2001), "The Autism Spectrum Quotient (AQ): Evidence from Asperger syndrome/high-functioning autism, males and females, scientists, and mathematicians," *Journal of Autism and Developmental Disorders* 31, 5–17.

11. On autism-related traits in students, see S. Wheelwright et al. (2006), "Predicting Autism Spectrum Quotient (AQ) from the Systemizing Quotient–Revised (SQ-R) and Empathy Quotient (EQ)," *Brain Research*, 1079, 47–56. On mathematical talent and autism, see S. Baron-Cohen et al. (2007), "Mathematical talent is linked to autism," *Human Nature* 18, 125–131.

12. See E. Ruzich et al. (2015), "Sex and STEM occupation predict Autism Spectrum Quotient (AQ) scores in half a million people," *PLoS ONE* 10, e0141229; Greenberg et al., "Testing the Empathizing-Systemizing (E-S) theory of sex differences."

13. The inverse correlation between the EQ and SQ is about −0.2 and is statistically significant. See Greenberg et al., "Testing the Empathizing-Systemizing (E-S) theory of sex differences."

14. On fetal testosterone, see S. Baron-Cohen et al. (2011), "Why are autism spectrum conditions more prevalent in males?," *PLoS Biology* 9, e1001081; S. Baron-Cohen et al. (2004), *Prenatal testosterone in mind: Amniotic fluid studies* (Cambridge, MA: MIT Press/Bradford Books); and M. M. McCarthy et al. (2015), "Surprising origins of sex differences in the brain," *Hormones and Behavior* 76, 3–10.

15. "Masculinization" simply refers to autistic people showing a shift toward having a profile that is usually more common in males. It is a contentious term, and we discuss the potential misunderstanding around this term in S. Baron-Cohen et al. (2018), "Autistic people do not lack empathy and nor are they hyper-male," *The Conversation*, November 12.

16. On early human neural sex differences, see J. Gilmore et al. (2018), "Imaging structural and functional brain development in early childhood," *Nature Reviews Neuroscience* 19(3), 127–137; and R. Knickmeyer et al. (2000), "Impact of sex and gonadal steroids on neonatal brain structure," *Cerebral Cortex* 24, 2721–2731. Again, these studies control for birthweight to ensure that this is not just a by-product of average differences in the size of male and female bodies.

17. See S. Baron-Cohen et al., *Prenatal testosterone in mind.*

18. On the role of prenatal testosterone and empathy and systemizing, see E. Chapman et al. (2006), "Fetal testosterone and empathy: Evidence from the Empathy Quotient (EQ) and the 'Reading the Mind in the Eyes' test," *Social Neuroscience* 1, 135–148; B. Auyeung et al. (2012), "Effects of fetal testosterone on visuospatial ability," *Archives of Sexual Behavior* 41, 571–581; and B. Auyeung et al. (2006), "Fetal testosterone and the Child Systemizing Quotient (SQ-C)," *European Journal of Endocrinology* 155, 123–130.

19. On the role of prenatal testosterone in brain development, see M. Lombardo et al. (2012), "Fetal testosterone influences sexually dimorphic gray matter in the human brain," *Journal of Neuroscience* 32, 674–680; and M. Lombardo et al. (2012), "Fetal programming effects of testosterone on the reward system and behavioral approach tendencies in humans," *Biological Psychiatry* 72, 839–847.

20. On the role of prenatal testosterone in language development, see S. Lutchmaya et al. (2002), "Fetal testosterone and vocabulary size in 18- and 24-month-old infants," *Infant Behaviour and Development* 24, 418–424.

21. See B. Auyeung et al. (2010), "Fetal testosterone and autistic traits in 18- to 24-month-old children," *Molecular Autism* 1, 11; and B. Auyeung et al. (2009), "Fetal testosterone and autistic traits," *British Journal of Psychology* 100, 1–22.

22. On the role of prenatal testosterone and estrogen in autism, see S. Baron-Cohen et al. (2015), "Elevated fetal steroidogenic activity in autism," *Molecular Psychiatry* 20, 369–376; S. Baron-Cohen et al. (2019), "Foetal estrogens and autism," *Molecular Psychiatry* 20, 369–376. Note the elevation was across all sex steroid hormones measured, which included prenatal testosterone. Prenatal sex steroid hormones are likely not the only driver of brain type, but rather interact with genetic predisposition, as genes are also involved too. And of course both sets of biological factors interact with experience.

23. See S. Baron-Cohen et al. (2001), "The 'Reading the Mind in the Eyes' test revised version: A study with normal adults, and adults with Asperger syndrome or high-functioning autism," *Journal of Child Psychology and Psychiatry* 42, 241–252.

24. On the genetic association with the Eyes test, see V. Warrier et al. (2018), "Genome-wide meta-analysis of cognitive empathy: Heritability, and correlates with sex, neuropsychiatric conditions and cognition," *Molecular Psychiatry* 23, 1402–1409. Genome-wide association studies are currently the most powerful way to identify if scores on any trait measure are correlated with any common genetic variants. This method requires large populations to uncover if common genetic variants—each of which contributes only a tiny amount to a trait—together and in certain combinations relate to scores on that trait. The common variant associated with the Eyes test is called rs7641347. All common genetic variants start with "rs-"—short for reference SNP, where an SNP is a single nucleotide polymorphism or a part of a gene that comes in different forms in the population—followed by a unique string of numbers. This SNP is on chromosome 3 (3p.26.1—the numbers provide its unique location).

25. On the genetic association with the Empathy Quotient, see V. Warrier et al. (2018), "Genome-wide analyses of self-reported empathy: Correlations with autism, schizophrenia, and anorexia nervosa," *Translational Psychiatry* 8, article 35. The common variant is called rs4882760, in a gene on chromosome 12 (12q.24.32). I use the word "association" here because we cannot be sure that these genes directly cause differences in empathy. The relevant genes may, for example, simply be involved in aspects of brain function that influence empathy, or they may influence other genes or hormones that influence empathy. Correlation famously does not equal causation. But the involvement of genetics at all points to empathy having evolved.

26. On the genetic association with systemizing, see V. Warrier et al. (2019), "Social and non-social autism symptom and trait domains are genetically dissociable," *Communications Biology* 2, 328. The SNPs that were significantly associated with systemizing were rs4146336 (on chromosome 3), rs1559586 (on chromosome 18), and rs8005092 (on chromosome 14).

27. On satisficing, see H. Simon (1956), "Rational choice and the structure of the environment," *Psychological Review* 63(2), 129–138. For the heritability of satisficing, see G. Saad et al. (2019), "Are identical twins more similar in their decision making styles than their fraternal counterparts?," *Journal of Business Research,* April.

28. On sibling recurrence rates of autism, see S. Ozonoff et al. (2011), "Recurrence risk for autism spectrum disorders: A baby siblings research consortium study," *Pediatrics* 128(3), e488–e495; P. Szatmari et al. (2016), "Prospective longitudinal studies of infant siblings of children with autism: Lessons learned and future directions," *Journal of the American Academy of Child and Adolescent Psychiatry* 55(3), 179–187.

29. On known genetic associations with autism, see Simons Foundation, "SFARI gene," www.sfari.org/resource/sfari-gene/; V. Warrier and S. Baron-Cohen (2017), "The genetics of autism," in *Encyclopedia of Life Sciences,* 1–9. On

the rate of rare genetic mutations in autism, and for a review of twin studies of autism, see G. Huguet et al. (2016), "The genetics of autism spectrum disorders," in P. Sassone-Corsi and Y. Christen, eds., *A time for metabolism and hormones: Research and perspectives in endocrine interactions* (Springer). On the role of common genetic variants in autism, see J. Grove et al. (2019), "Identification of common genetic risk variants for autism spectrum disorder," *Nature Genetics* 51, 431–444.

30. On the genetic association between autism and mathematical ability, see S. Baron-Cohen et al. (2007), "Mathematical talent is linked to autism," *Human Nature* 18, 125–131.

31. See J. Monroe (2019), "Go, Greta: Autism is my superpower too," *Guardian*, April 27, www.theguardian.com/society/2019/apr/27/jack-monroe-autism-is -my-superpower-like-greta-thunberg; N. Prouix (2019), "Becoming Greta," *New York Times*, February 21; S. Baron-Cohen (2020), "Without such families speaking out, their crises remain hidden," part of L. Carpenter (2020), "Greta and Beata: How autism and climate activism affected the Thunberg family," *Times* (of London) *Magazine*, February 28; and G. Thunberg, tweet of August 31, 2019, twitter.com/GretaThunberg/status/1167916177927991296?ref_src=twsrc%5E google%7Ctwcamp%5Eserp%7Ctwgr%5Etweet.

CHAPTER 4: THE MIND OF AN INVENTOR

1. See D. Hajela (2008), "Scientists to capture DNA of trees worldwide for database," *USA Today*, May 2; and Botanic Gardens Conservation International (BGCI), "Global tree assessment," www.bgci.org/our-work/projects -and-case-studies/global-tree-assessment/.

2. There are several well-documented examples of people with hypermnesia, including Jill Price, Marilu Henner, and Aurelian Hayman. See G. Marcus (2009), "Total recall: The women who can't forget," *Wired*, March 23; *The boy who can't forget* (2012, documentary), September 25; and A. Ward (2012), "Total recall," *Sunday Times*, September 23. Note that hypermnesia is also sometimes referred to as hyperthymesia.

3. Jonah's talent is in semantic memory. Remembering where you were in a memory is called "autobiographical" memory, whereas remembering facts is called "semantic" memory.

4. On autistic adults living with their parents, see K. Anderson et al. (2013), "Prevalence and correlates of postsecondary residential status among young adults with an autism spectrum disorder," *Autism* 18(5), 562–570.

5. I participated in the BBC documentary *Employable me* about autistic people who, like Jonah, simply want a single chance to work or to prove themselves through unpaid work experience.

6. See S. Cassidy et al. (2014), "Suicidal ideation and suicide plans or attempts in adults with Asperger's syndrome attending a specialist diagnostic clinic: A clinical cohort study," *Lancet Psychiatry* 1, 142–147; S. Griffiths et al. (2019), "The Vulnerability Experiences Quotient (VEQ): A study of vulnerability, mental

health, and life satisfaction in autistic adults," *Autism Research* 10, 1516–1528. I discussed the urgent need for suicide prevention in the autism community in my speech at the United Nations in 2017: see S. Baron-Cohen (2017), "Toward autonomy and self-determination," UN speech on Autism Awareness Day 2017, March 31, webtv.un.org/meetings-events/watch/toward-autonomy-and-self -determination-world-autism-awareness-day-2017/5380816054001.

7. On autism and emotion recognition, see O. Golan et al. (2006), "The Cambridge Mindreading (CAM) Face-Voice Battery: Testing complex emotion recognition in adults with and without Asperger syndrome," *Journal of Autism and Developmental Disorders* 36, 169–183; O. Golan and S. Baron-Cohen (2006), "Systemizing empathy: Teaching adults with Asperger syndrome or high functioning autism to recognize complex emotions using interactive multimedia," *Development and Psychopathology* 18, 591–617; and O. Golan et al. (2006), "The 'Reading the Mind in Films' task: Complex emotion recognition in adults with and without autism spectrum conditions," *Social Neuroscience* 1, 111–123.

8. See S. Baron-Cohen (1997), "Hey! It was just a joke! Understanding propositions and propositional attitudes by normally developing children, and children with autism," *Israel Journal of Psychiatry* 34, 174–178.

9. On feeling as if one is from a different planet, see Wrong Planet, www .wrongplanet.net; and C. Sainsbury (2009), *Martian in the playground: Understanding the school child with Asperger syndrome* (London: Sage Publications). Many autistic children identify with the title of Clare Sainsbury's book. Clare's foundation, Three Guineas Trust, kindly funded our clinic for the late diagnosis of Asperger syndrome from 1997 through 2010, at a time when the National Health Service said that diagnosing adults was not a priority. See S. Baron-Cohen (2007), "The lost generation," *Communication* 41, 12–13; and M. C. Lai and S. Baron-Cohen (2015), "Identifying the lost generation," *Lancet Psychiatry*, November.

10. On autism and cognitive empathy, see S. Baron-Cohen et al. (2014), "Attenuation of typical sex difference in 800 adults with autism vs. 3,900 controls," *PLoS ONE* 9, e102251; Greenberg et al., "Testing the Empathizing-Systemizing (E-S) theory of sex differences"; and Baron-Cohen (2011), *Zero degrees of empathy*.

11. On autism and affective empathy, see P. Rueda et al. (2015), "Dissociation between cognitive and affective empathy in youth with Asperger syndrome," *European Journal of Developmental Psychology* 12, 85–98; and S. Baron-Cohen (2013), "Empathy deficits in autism and psychopaths: Mirror opposites?," in M. Banaji and S. Gelman, eds., *Navigating the social world: What infants, children, and other species can teach us* (Oxford: Oxford University Press).

12. On Daniel Tammet, see S. Baron-Cohen et al. (2007), "Savant memory in a man with colour form-number synaesthesia and Asperger syndrome," *Journal of Consciousness Studies* 14, 237–251; and D. Bor et al. (2007), "Savant memory for digits in a case of synaesthesia and Asperger syndrome is related to hyperactivity in the lateral prefrontal cortex," *Neurocase* 13, 311–319. Daniel is the author of many books, including D. Tammet (2006), *Born on a blue day: Inside the extraordinary mind of an autistic savant* (London: Hodder & Stoughton). You

can watch the episode that includes my diagnostic interview with him at "Savant learns to speak Icelandic in a week," YouTube, July 9, 2012, www.youtube.com /watch?v=_GXjPEkDfek.

13. The American Psychiatric Association (APA), which publishes the *Diagnostic and Statistical Manual* (*DSM*), decided that the fifth edition (*DSM-5*, 2013) would drop the term "Asperger syndrome," for reasons to do with the unreliability of the use of the term (or poor agreement) among clinicians. I was one of those who spoke out as early as 2009 against throwing out this term because ultimately we do need to identify subgroups in the otherwise very broad spectrum of autism. See S. Baron-Cohen (2009), "The short life of a diagnosis," op-ed in *New York Times*, November 10. The reliability issue could have been resolved by tightening up the definition rather than abandoning the term altogether. Our clinic in Cambridge continued using the term "Asperger syndrome" until 2018 after the revelation that Hans Asperger, the pediatrician who was the term's namesake, collaborated with the Nazis. See H. Czech (2018), "Hans Asperger, National Socialism, and 'race hygiene' in Nazi-era Vienna," *Molecular Autism* 9, 29; and S. Baron-Cohen et al. (2018), "Did Hans Asperger actively assist the Nazi euthanasia program?," *Molecular Autism* 9. The World Health Organization, which publishes the International Classification of Diseases (ICD), decided in its eleventh edition in 2019 to stop using the term too. Most clinicians and scientists now use a single umbrella term of "autism" (or autism spectrum disorder), which is very broad. I have argued for the need for subgrouping. See S. Baron-Cohen (2018), "Is it time to give up on a single diagnostic label for autism," *Scientific American*, May 4, https://blogs.scientificamerican.com/observations/is -it-time-to-give-up-on-a-single-diagnostic-label-for-autism/.

14. On sensory hypersensitivity and autism, see T. Tavassoli et al. (2014), "The Sensory Perception Quotient (SPQ): Development and validation of a new sensory questionnaire for adults with and without autism," *Molecular Autism* 5(29).

15. On Lauri Love, see O. Bowcott and D. Taylor (2016), "Hacking suspect could kill himself if extradited to the US, court told," *Guardian*, June 28. Gary McKinnon was another autistic "ethical hacker" I diagnosed while he was facing similar charges in 2008, for hacking into the Pentagon (described as the "biggest military computer hack of all time"). He was arrested in 2002 and had to wait ten years for a decision about extradition. Like Lauri, Gary said that he would prefer to die by suicide rather than endure the brutality of an American jail. He told me that he had systemized how he could kill himself so that no police force would be able to detect his means or prevent him from doing so. Thankfully, in 2012, Gary got the same good news Lauri eventually did: he would not be extradited. See S. Marsden (2009), "Hacker was naive not criminal says expert," *Independent*, January 15.

16. See Malcolm Cowley's definition of "genius" in his introduction to the novel *Anna Karenina* (1877): "Genius is vision, often involving the gift of finding patterns where others see nothing but a chance collection of objects." L. Tolstoy (1960), *Anna Karenina* (New York: Bantam); see also E. Anderson (2013), "Three things you can do to think like a genius," *Forbes*, January 7.

17. The following account is based on Beals, "The biography of Thomas Edison." See also J. Gernter (2013), *The idea factory: Bell Labs and the great age of American innovation* (New York: Penguin Books).

18. As quoted in J. L. Elkhorne (1967), "Edison—The fabulous drone," in *73* 46(3), 52. The quote from Edison is disputed but not disproven. See Wikipedia, "Thomas Edison," en.wikiquote.org/wiki/Thomas_Edison #Disputed. The idea that Edison may have been autistic is suggested in several blogs, such as Applied Behavior Analysis, "History's 30 most inspiring people on the autism spectrum," www.appliedbehavioranalysisprograms.com /historys-30-most-inspiring-people-on-the-autism-spectrum/.

19. On Nikola Tesla, see C. Eldrid-Cohen (2016), "Historical figures who may have been on the autism spectrum," The Art of Autism, October 20, the-art -of-autism.com/historical-figures-who-may-have-been-on-the-autism-spectrum/; and "Was Nikola Tesla autistic?," AppliedBehaviorAnalysisEdu.org, www.applied behavioranalysisedu.org/was-nikola-tesla-autistic/. Note that although the latter website is related to applied behavioral analysis, this is not relevant to Tesla's biography.

20. See P. Galanes (2018), "The mind meld of Bill Gates and Steven Pinker," *New York Times*, January 27; and S. Levy (2019), "*Inside Bill's Brain* calls BS on Malcolm Gladwell's outliers theory," *Wired*, September 20.

21. The sigma character is taken from Wikipedia, "The common Six Sigma symbol," en.wikipedia.org/wiki/Six_Sigma#/media/File:Six_sigma-2.svg; see also D. Dusharme, "Six sigma survey: Breaking through the six sigma hype," *Quality Digest*, www.qualitydigest.com/nov01/html/sixsigmaarticle.html. The concept of *Six sigma* is not without its critics but is nevertheless now a powerful byword for making things work to an awesome level of quality.

22. On engineers, see G. Madhavan (2015), *Applied minds: How engineers think* (London: W. W. Norton and Co. Ltd.).

23. This phrase comes from Madhavan (2015), *Applied minds*.

24. The statistic on plane crashes comes from "How many airplanes take off each hour on average in the world?," Quora, www.quora.com/How-many-air planes-take-off-each-hour-on-average-in-the-world; see also B. Bowman (2017), "How do people survive plane crashes?," *Curiosity*, August 2; FlightAware, uk.flight aware.com/live/; and L. Smith-Spark (2019), "Plane crash deaths rise in 2018 but accidents are still rare," CNN, January 3, edition.cnn.com/2019/01/02 /health/plane-crash-deaths-intl/index.html. This example of plane crashes as a perspective on engineering is discussed in Madhavan (2015), *Applied minds*.

25. See V. Cerf (2009), "The day the internet age began," *Nature* 461(7268), 1202–1203. Madhavan (2015), *Applied minds*, discusses the peppercorn grinder example to illustrate what he calls the "emergence"—nothing about a single peppercorn would cause congestion in the system, beyond knowing that it has the property of friction. But rate of flow—a higher-order property, not the basic units (peppercorns) themselves—can lead to the emergent property of congestion.

26. President John F. Kennedy made his historic speech on September 12, 1962, announcing that the United States was going to send a rocket to the

moon—a mere 240,000 miles away—and return it safely to the United States. Engineers faced at least four huge challenges: metal alloys withstanding the heat and stress of flying at high speed through the atmosphere; carrying the equipment it needed for propulsion, guidance, control, communications, food, and survival; an unknown landing target (no one had ever been there before); and reentering the atmosphere at 25,000 miles per hour, which would heat up the rocket to a temperature half that of the sun. Yet engineers systemized the problem to find a solution.

27. See S. Wade-Leeuwen et al. (2018), "What's the difference between STEM and STEAM?," *The Conversation,* June 10.

28. On Glenn Gould, see K. Bazzana et al. (2008), "Glenn Gould," in *The Canadian encyclopedia* (Historica Canada), August 17, 2008, updated March 4, 2015; L. McLaren (2000), "Was Glenn Gould autistic?," *Globe and Mail,* February 1; P. Ostwald (1997), *Glenn Gould: The ecstasy and tragedy of genius* (New York: W. W. Norton & Co.); and K. Bazzana (2003), *Wondrous strange: The life and art of Glenn Gould* (Toronto: McClelland & Stewart). The quote from his father comes from the documentary *Glenn Gould: A portrait,* by E. Till and V. Tovell (1985), www.youtube.com/watch?v=fV3IdRGJ-Bk.

29. See Jonathan Chase's excellent TEDx talk, "Music as a window into the autistic mind," November 17, 2014, www.youtube.com/watch?v=MxxUhW7d8yI&feature=share.

30. Psychologist David Greenberg studied if systemizers and empathizers prefer different styles of music. He found that systemizers (those who scored higher on the SQ than the EQ) prefer more "intense" music (punk, heavy metal, hard-rock genres), high arousal (strong, tense, thrilling attributes), positive valence (animated), and cerebral depth (complexity). In contrast, empathizers (those who scored higher on the EQ than the SQ) preferred more "mellow" music (R&B/soul, soft rock, adult contemporary genres), low arousal (gentle, warm, sensual attributes), negative valence (depressing and sad), and emotional depth (poetic, relaxing, and thoughtful). So, systemizing and empathy pervade how we look at and listen to every aspect of the world. See D. M. Greenberg et al. (2015), "Musical preferences are linked to cognitive styles," *PLoS ONE* 10(7), e0131151.

31. On autism and the Rubik's cube, see S. Baron-Cohen et al. (2009), "Talent in autism: Hyper-systemizing, hyper-attention to detail, and sensory hyper-sensitivity," *Proceedings of the Royal Society, Philosophical Transactions: Series B* 364, 1377–1383. On Max Park, see J. Rapson (2017), "They said autism meant he'd need life-long care—then he got a Rubik's cube," For Every Mum, July 29, foreverymom.com/family-parenting/autism-rubiks-cube-max-park/. On June 18, 2014, we hosted an event in Cambridge Union called "Autism and the Rubik's Cube: Creating order from chaos," with Professor Ernesto Rubik (architect and inventor of the cube); see "Event investigates 'Autism and the Rubik's Cube: Creating order from chaos,'" Cambridge Network, June 24, 2014, www.cambridgenetwork.co.uk/news/event-investigates-autism-and-the-rubiks-cube.

32. On Kobe Bryant, see A. Tsuji (2016), "Jamal Crawford adds to the list of legendary Kobe Bryant practice stories," *USA Today,* January 28.

33. On whether Andy Warhol was autistic, see M. Fitzgerald (2014), "Andy Warhol and Konrad Lorenz: Two persons with Asperger's syndrome," professormichaelfitzgerald.eu/andy-warhol-and-konrad-lorenz-two-persons-with-aspergers-syndrome/. On whether Wittgenstein was autistic, see S. Wolf (1995), *Loners: The life path of unusual children* (East Sussex, UK: Psychology Press); and M. Fitzgerald (2004), *Autism and creativity: Is there a link between autism in men and exceptional ability?* (East Sussex, UK: Brunner-Routledge). On whether Hans Christian Andersen was autistic, see J. Brown (2007), "Ice puzzles of the mind: Autism and the writings of Hans Christian Andersen," *CEA Critic* 69(3), 44–64. On whether Albert Einstein was autistic, see N. Fleming (2008), "Albert Einstein 'found genius through autism,'" *Telegraph*, February 21, www.telegraph.co.uk /news/science/science-news/3326317/Albert-Einstein-found-genius-through -autism.html. On whether Henry Cavendish was autistic, see S. Silberman (2015), *Neurotribes: The legacy of autism and how to think smarter about people who think differently* (New York: Penguin Random House). See also I. James (2003), "Singular scientists," *Journal of the Royal Society of Medicine*, 96, 36–39. Finally, the Autism Community Network adds the names of famous people to a list of those who may have been autistic at www.autismcommunity.org.au/famous—with-autism.html.

CHAPTER 5: A REVOLUTION IN THE BRAIN

1. Associative learning includes classical conditioning, where an animal learns that A is paired with B and B is a reward or a punishment; and operant conditioning, where an animal learns that doing action A leads to B and B, again, is a reward or a punishment. Associative learning is seen in many species, from rats to humans. See Skinner (1938), *Behavior of organisms.*

2. There is some debate about whether *Homo habilis* was the first hominid to use tools, as *Australopithecus garhi*, who lived 2.6 million years ago, has been found with some stone tools, and this would have been 100,000 to 200,000 years earlier than *Homo habilis*. See S. Oppenheimer (2004), *The real Eve: Modern man's journey out of Africa* (New York: Carroll and Graf). For a useful online resource from Oppenheimer's research group, including a discussion on whether any of the other gracile Australopithecines used stone tools, see Bradshaw Foundation, www.bradshawfoundation.com.

3. The main change in the tool-making of *Homo erectus* was that they made their axes symmetrically; working on both sides, they produced a tool that allowed for more uses, such as slicing. They also used tools to extract marrow from bones. *Homo erectus* may have been the first species to bury their dead, suggesting, if true, that they could contemplate mortality. I think this would require a lot more evidence. For the Smithsonian National Museum of Natural History's useful online resource discussion of this, see "The mystery of the pit of bones, Atapuerca, Spain," humanorigins.si.edu/research/whats-hot-human-origins /mystery-pit-bones-atapuerca-spain. See also I. de la Torre and S. Hirata (2015), "Percussive technology and human evolution: An introduction to a comparative approach in fossil and living primates," *Philosophical Transactions of the Royal*

Society: Series B 370, 20140346; and B. Pobiner (2013), "Evidence for meat-eating by early humans," *Nature Education Knowledge* 4(6), 1; and Y. Harari (2015), *Sapiens: A brief history of humankind* (New York: HarperCollins).

4. Some authorities suggest that Neanderthals began as far back as 400,000 years ago, but a cautious date is from 300,000 years ago. See M. Marshall (2012), "Neanderthals were ancient mariners," *New Scientist*, March 3, 2854; J. Shea (2003), "Neanderthals, competition, and the origin of modern human behavior in the Levant," *Evolutionary Anthropology* 12, 173–187; A. Sorensen et al. (2018), "Neandertal fire-making technology inferred from microwear analysis," *Scientific Reports* 8(1), 10065; and C. M. Turcotte (n.d.), "Exploring the fossil record: Stone tools," Bradshaw Foundation, www.bradshawfoundation.com/origins /mousterian_stone_tools.php.

5. On the controversy over Neanderthal adhesives, see P. Kozowyk and J. Poulis (2019), "A new experimental methodology for assessing adhesive properties shows that Neandertals used the most suitable material available," *Journal of Human Evolution* 137, 102664; I. Degano et al. (2019), "Hafting of Middle Paleolithic tools in Latium (central Italy): New data from Fossellone and Sant'Agostino caves," *PLoS ONE* (June 20); and M. Niekus et al. (2019), "Middle Paleolithic complex technology and a Neandertal tar-backed tool from the Dutch North Sea," *Proceedings of the National Academy of Sciences* 116(44), 22081–22087. Other archaeologists challenge this view. See P. Schmidt et al. (2019), "Birch tar production does not prove Neanderthal behavioral complexity," *Proceedings of the National Academy of Sciences* 116(36), 17707–17711. On the controversy over Neanderthal intentional burial, see H. Dibble et al. (2015), "A critical look at evidence from La Chapelle-aux-Saints supporting an intentional burial," *Journal of Archaeological Science* 53, 649–657.

6. On *Homo sapiens*, see I. Herschkovitz et al. (2018), "The earliest modern humans outside Africa," *Science* 359, 456–459; T. White et al. (2003), "Pleistocene *Homo sapiens* from Middle Awash, Ethiopia," *Nature* 423, 742–747; and C. Henshilwood et al. (2002), "Emergence of modern human behavior: Middle Stone Age engravings from South Africa," *Science* 295, 1278–1280.

7. The oldest perforated shells are from North Africa 82,000 years ago. See A. Bouzouggar et al. (2007), "82,000-year-old shell beads from North Africa and implications for the origins of modern human behaviour," *Proceedings of the National Academy of Sciences* 104(24), 9964–9969. For a report on the slightly younger perforated shells in Blombos Cave in South Africa (from 75,000 years ago), see C. Henshilwood et al. (2004), "Middle Stone Age shell beads from South Africa," *Science* 304, 404. There are multiple examples of jewelry-making from around this time. Much clearer examples of jewelry are found from 40,000 years ago, made from ostrich eggshells in East Africa, and from mammoth ivory from 42,000 years ago in Swabia. See also C. Henshilwood and K. van Niekerk (2016), "What excavated beads tell us about the when and where of human evolution," *The Conversation*, January 28. To see the seven oldest pieces of jewelry ever found, see Ancient Facts, "7 oldest pieces of jewelry in the world," www.ancientfacts.net/7-oldest-pieces-jewelry-world/. Polycentric

(multiple holes) rather than monocentric (one central hole) jewelry is particularly complex.

8. On the bow-and-arrow, see K. S. Brown et al. (2011), "An early and enduring advanced technology originating 71,000 years ago in South Africa," *Nature* 491, 590–593; M. Lombard (2011), "Quartz-tipped arrows older than 60 ka: Further use-trace evidence from Sibudu, Kwa-Zulu-Natal, South Africa," *Journal of Archaeological Science* 38(8), 1918–1930; and M. Lombard and L. Phillips (2010), "Indications of bow and stone-tipped arrow use 64,000 years ago in KwaZulu-Natal, South Africa," *Antiquity* 84(325): 635–648.

9. In South Africa we see blade technology, dating from about 100,000 years ago—the so-called Fauresmith tools. The most generous interpretation of these new tools is that humans had started to make *specialized* tools, and a blade would certainly meet this definition. Some question if these tools were a significant advance. For example, there are examples of blade technologies before 100,000 years ago in southern Africa, and there are examples among Neanderthals in Europe and archaic hominins in the Near East. There are also 300,000-year-old blade assemblages in the Middle Stone Age in Kenya. See A. Herries (2011), "A chronological perspective on the Acheulian and its transition to the Middle Stone Age in Southern Africa: The question of the Fauresmith," *International Journal of Evolutionary Biology* article 961401, 1–25; and D. Underhill (2011), "The study of the Fauresmith: A review," *South African Archaeological Bulletin* 66(193), 15–26.

10. On the earliest engravings, see C. Henshilwood et al. (2002), "Emergence of modern human behavior"; D. Perlman (2002), "Cave's ancient treasure: 77,000-year-old artifacts could mean human culture began in Africa," SFGate, January 11, www.sfgate.com/news/article/Cave-s-ancient-treasure-77-000-year-old-2883686.php; and P. J. Texier et al. (2010), "A Howiesons Poort tradition of engraving ostrich eggshell containers dated to 60,000 years ago at Diepkloof Rock Shelter, South Africa," *Proceedings of the National Academy of Sciences* 107(14), 1680–1685. Some engravings as old as 500,000 years have been reported, suggesting that engraving may date back to *Homo erectus*, but interpretation of these one-off examples is controversial. See J. Joordens et al. (2015), "*Homo erectus* at Trinil on Java used shells for tool production and engraving," *Nature* 518, 228–231; and H. Thompson (2014), "Zigzags on a shell from Java are the oldest human engravings," *Smithsonian*, December 3, www.smithsonianmag.com/science-nature/oldest-engraving-shell-tools-zigzags-art-java-indonesia-humans-180953522/. Although there is a one-off example of possible engraving from Java 500,000 years ago, made by *Homo erectus*, this must currently be interpreted with caution, as there are no other pieces of evidence to support the idea that they could make systematic variations in design.

11. On the first boats, see V. Macauley (2005), "Single, rapid coastal settlement of Asia revealed by analysis of complete mitochondrial genomes," *Science* 308(5724), 1034–1036; A. Thorne et al. (1999), "Australia's oldest human remains: Age of the Lake Mungo 3 skeleton," *Journal of Human Evolution* 36(6), 591–612; J. O'Connell and J. Allen (1998), "When did humans first arrive in

Greater Australia and why is it important to know?," *Evolutionary Anthropology* 6(4), 132–146; and R. Bednarik (2003), "Seafaring in the Pleistocene," *Cambridge Archaeological Journal* 13(1), 41–66.

12. On the earliest fishing, see S. O'Connor et al. (2011), "Pelagic fishing at 42,000 years before the present and the maritime skills of modern humans," *Science* 334, 1117–1121; and Z. Corbyn (2011), "Archaeologists land world's oldest fishing hook," *Nature*, November 24. The existence of the earliest fishing hook receives converging evidence from an analysis of the bones of a man found in China from 40,000 years ago, confirming that his diet included freshwater fish. See Y. Yaowu Hu et al. (2009), "Stable isotope dietary analysis of the Tianyuan 1 early modern human," *Proceedings of the National Academy of Sciences* 106(27), 10971–10974; and M. Price (2016), "World's oldest fishing hook found in Okinawa," *Science*, September 16.

13. On the first adorned graves, see G. Giacobini (2007), "Richness and diversity of burial rituals in the Upper Paleolithic," *Diogenes* 54(2), 19–39. On cave art, see M. Aubert et al. (2014), "Pleistocene cave art from Sulawesi, Indonesia," *Nature* 514, 223–227.

14. On the first dwellings, see J. Kolen (2000), "Hominids without homes: On the nature of Middle Paleolithic settlement in Europe," in W. Roebroeks et al., eds., *The Middle Paleolithic occupation of Europe* (Leiden: Leiden University Press); and P. Mellars (1996), *The Neanderthal legacy: An archaeological perspective from Western Europe* (Princeton, NJ: Princeton University Press).

15. On bone needles, see L. Backwell et al. (2008), "Stone Age bone tools from the Howiesons Poort layers, Sibudu Cave, South Africa," *Journal of Archaeological Science* 35(6), 1566–1580; and M. Collard et al. (2016), "Faunal evidence for a difference in clothing use between Neanderthals and early modern humans in Europe," *Journal of Anthropological Archeology* 44B, 235–246.

16. Claims that jewelry was made by Neanderthals have been questioned: did they make jewelry or were they given jewelry by humans? See L. Geggel (2016), "Neanderthals fashioned jewelry out of animal teeth and shells," *Live Science*, September 27; E. Calloway (2014), "Neanderthals made some of Europe's oldest art," *Nature*, September 1; and O. Rudgard (2018), "Neanderthal art was far better than previously thought as scientists find they made earliest cave paintings," *Daily Telegraph*, February 22. The same controversy surrounds the existence of eagle claws and shells: were these made or perforated by Neanderthals and were they worn as jewelry? See E. Calloway (2015), "Neanderthals wore eagle talons as jewelry," *Nature*, March 11; D. Hoffman et al. (2018), "Symbolic use of marine shells and mineral pigments by Iberian Neandertals 115,000 years ago," *Science Advances* 4(2), 5255; S. McBrearty and A. Brooks (2000), "The revolution that wasn't: A new interpretation of the origin of modern human behavior," *Journal of Human Evolution* 39, 453–563; E. Yong (2018), "A cultural leap at the dawn of humanity," *Atlantic*, March 15; R. Becker (2018), "Ancient cave paintings turn out to be by Neanderthals, not modern humans," *The Verge*, February 22; J. Rodrigues-Vidal et al. (2014), "A rock engraving made by Neanderthals in Gibraltar," *Proceedings of the National Academy of Sciences* 111(37), 13301–13306;

E. Calloway (2014), "Neanderthals made some of Europe's oldest art," *Nature,* September 1; and, most recently, R. White et al. (2019), "Still no archaeological evidence that Neanderthals created Iberian cave art," *Journal of Human Evolution* (October).

17. Composite figures like the lion-man sculpture found in Germany are also seen in a bird-man in Lascaux, a lion-woman from Chauvet, and the engraving of a bird-horse man from Hornos Le Pena, all around 37,000 years ago. And in Czechoslovakia 29,000 years ago, humans were shaping similar "Venus figures" of women in ceramic, from clay. See R. Dalton (2003), "Lion man takes pride of place as oldest statue," *Nature* 425, 7; J. Wilford (2009), "Full-figured statuette, 35,000 years old, provides new clues to how art evolved," *New York Times,* May 13; R. Lesure (2002), "The Goddess diffracted: Thinking about the figurines of early villages," *Current Anthropology* 43(4), 587–610; and R. White (2006), "The women of Brassempouy: A century of research and interpretation," *Journal of Archaeological Method and Theory* 13 (December), 4. This archaeological tradition falls within the Aurignacian period, from 43,000 years ago to 26,000 years ago. See also N. Conard (2015), "Cultural evolution during the middle and late Pleistocene in Africa and Eurasia," in W. Henke and I. Tattersall, eds., *Handbook of paleoanthropology* (Berlin: Springer-Verlag); and N. Conard (2009), "A female figurine from the basal Aurignacian of Hohle Fels Cave in southwestern Germany," *Nature* 459, 248–252.

18. The Lower Paleolithic period is defined as when stone tools first appeared 3.3 million years ago. The beginning of the Middle Paleolithic period is defined as when Mousterian tools are first observed approximately 300,000 years ago. The Upper Paleolithic period is dated from about 50,000 years ago, with the appearance of the clear changes in complexity of tools and the evidence of art that I and others refer to as the cognitive revolution. I date this cognitive revolution to slightly earlier (70,000 to 100,000 years ago). The Upper Paleolithic period ended approximately 10,000 years ago (with the beginning of the current geological era, the Holocene, when technology and civilizations really took off).

19. In suggesting that the cognitive revolution occurred more like 40,000 to 50,000 years ago, archaeologist Richard Klein opts for the more recent time window because the evidence is much clearer. See R. Klein and B. Edgar (1992), *The dawn of human culture* (New York: Vintage).

20. On the cognitive revolution, see R. Klein (2017), "Language and human evolution," *Journal of Neurolinguistics* 43(B), 204–221; Klein and Edgar, *The dawn of human culture;* and N. Conard (2010), "Cultural modernity: Consensus or conundrum?," *Proceedings of the National Academy of Sciences* 107(17), 7621–7622.

21. On an abrupt versus incremental evolution of "new thinking" in human cognition, see C. Hayes (2012), "New thinking: The evolution of human cognition," *Philosophical Transactions of the Royal Society of London: Series B* 367, 2091–2096. This paper is included in a thematic collection of articles, *Philosophical Transactions of the Royal Society of London: Series B,* ed. U. Frith and C. Hayes, 367(1599, 2012), 1471–2970. Hayes challenges the idea of abrupt cognitive change between humans and our ancestors, in favor of incremental changes.

22. See S. López et al. (2015), "Human dispersal out of Africa: A lasting debate," *Evolutionary Bioinformatics* 11, 57–68; and N. Conard (2008), "A critical view of the evidence for a Southern African origin of behavioural modernity," *South African Archaeological Society Goodwin Series* 10, 175–178.

23. Reports at the time dated the bone flute to be at least 35,000 years old, and Nicholas Conard wrote in an email to *New York Times* reporter John Noble Wilford that it was more like 40,000 years old. The flute is in the Urgeschichtliches Museum. See J. N. Wilford (2009), "Flutes offer clues to Stone-Age music," *New York Times*, June 24, www.nytimes.com/2009/06/25/science/25flute.html. The flute from Hohle Fels, found in 2008, at the time was reported to be the oldest flute in the world. It was thought to be older than the one made from a swan and the other made from a woolly mammoth ivory found in the Geissenklösterle cave nearby. In 2012, these flutes were carbon-dated to actually be 42,000 years old. See T. Higham et al. (2012), "Testing models for the beginnings of the Aurignacian and the advent of figurative art and music: The radiocarbon chronology of Geissenklösterle," *Journal of Human Evolution* 62(6, May 8), 664–676. The oldest bone flute was found in the Geissenklösterle cave in southern Germany, made from the wing bones of large birds—a swan in the best-preserved example. A larger set of bone flutes discovered in 1921 in the Pyrenees dates to between 27,000 and 20,000 years old. See N. Conard et al. (2009), "New flutes document the earliest musical tradition in southwestern Germany," *Nature* 460(7256), 737–740. The four caves in southern Germany—Vogelherd, Hohlenstein-Stadel, Geissenklösterle, and Hohle Fels—are all close to one another. Sadly, tools made of wood or other degradable materials would have long since vanished from the archaeological record.

24. The four holes are clearly visible in the photo in figure 5.8, while the fifth one is partly visible where the flute has been broken. The rest of the flute has not been recovered.

25. See J. Powell (2010), *How music works: The science and psychology of beautiful sounds, from Beethoven to the Beatles and beyond* (New York: Little, Brown and Co.). I have assumed that a generation is 25 years, so 40,000 years is 1,600 generations. See A. Ockelford (2018), *Comparing notes: How we make sense of music* (London: Profile Books). Ockelford argues that musical appreciation involves not only recognizing patterns but putting yourself into the mind of the music-maker, to imagine how they were trying to vary the pattern in intentional ways. In this sense, music appreciation involves both the Systemizing Mechanism and the Empathy Circuit. But a person with reduced empathy could still appreciate music by feeding it through their Systemizing Mechanism, focusing just on the if-and-then patterns.

26. On the reward circuitry in the human brain while listening to music, see V. Salimpoor et al. (2011), "Anatomically distinct dopamine release during anticipation and experience of peak emotion to music," *Nature Neuroscience* 14, 257–262; and D. Västfjäll (2001), "Emotion induction through music: A review of the musical mood induction procedure," *Musicae Scientiae* 5(1, suppl.), 173–211.

27. On the reward circuitry in the bird brain after hearing birdsong, see S. Earp and D. Maney (2012), "Birdsong: Is it music to their ears?," *Frontiers of Evolutionary Neuroscience*, November 28.

28. On whether birds recognize rhythm and music, see C. Cate et al. (2016), "Can birds perceive rhythmic patterns? A review of experiments on a song bird and a parrot species," *Frontiers of Psychology*, May 19. Marcelo Araya-Salas analyzed the nightingale's birdsong and discovered it is not harmonic, in the sense that the intervals between adjacent notes in birdsong are not in consistent relationships with each other. See M. Araya-Salas (2012), "Is birdsong music? Evaluating harmonic intervals in songs of a neotropical songbird," *Animal Behaviour* 84(2), 309–313; E. Underwood (2012), "Birdsong not music after all," *Science*, August 15; and M. Araki et al. (2016), "Mind the gap: Neural coding of species identity in birdsong prosody," *Science* 354(6317), 1282–1287.

29. See V. Dufour et al. (2015), "Chimpanzee drumming: A spontaneous performance with characteristics of human musical drumming," *Scientific Reports* 5, 11320.

30. See S. Kirschner and M. Tomasello (2009), "Joint drumming social context facilitates synchronization in preschool children," *Journal of Experimental Child Psychology* 102, 299–314.

31. See C. Snowdon and D. Teie (2009), "Affective responses in Tamarins elicited by species-specific music," *Biology Letters* 6, 30–32; and A. Patel (2014), "The evolutionary biology of musical rhythm: Was Darwin wrong?," *PLOS Biology* 12(3), e1001821.

32. See S. Coren (2012), "Do dogs have a musical sense?," *Psychology Today*, April 2; A. Bowman et al. (2017), "The effect of different genres of music on the stress levels of kennelled dogs," *Physiology and Behavior* 171(15), 207–215; and A. Bowman et al. (2015), "'Four Seasons' in an animal rescue centre: Classical music reduces environmental stress in kennelled dogs," *Physiology and Behavior* 143(15), 70–82.

33. There is no evidence that Neanderthals had rhythm perception or made musical instruments. See S. Mithen (2005), *The singing Neanderthals* (Cambridge, MA: Harvard University Press); and F. D'Errico et al. (1998), "A Middle Paleolithic origin of music? Using cave bear bone accumulations to assess the Divje Babe bone 'flute,'" *Antiquity* 72, 65–79.

34. Charles Darwin proposed a different theory of music: he thought that music in humans served an evolutionary function as part of courtship and sexual selection. See C. Darwin (1871), *The descent of man, and selection in relation to sex* (London: John Murray); and see P. Kivy (1959), "Charles Darwin on music," *Journal of the American Musicological Society* 12(1), 42–48. Others have argued that music's evolutionary function is to enhance social group cohesion. See E. Hagen and G. Bryant (2003), "Music and dance as a coalition signalling system," *Human Nature* 14(1), 21–51. While both of these claims are likely to be true, a much more basic question is how we even detect musical patterns. I argue that we do so because we are *if-and-then* pattern-seekers. Subsequently, both natural selection and we ourselves have co-opted music into a great many human social activities.

35. F. d'Errico et al. (2012), "Early evidence of San material culture represented by organic artifacts from Border Cave, South Africa," *Proceedings of the National Academy of Sciences* 109(33), 13214–13219. The original report is published in P. Beaumont (1973), "Border cave—A progress report," *South African Journal of Science* 69, 41–46. Note that there are even older "notched" bones dated back to 80,000 years, but the meaning of the notches is less clear. See R. Vogelsang et al. (2010), "New excavations of Middle Stone Age deposits at Apollo 11 Rockshelter, Namibia: Stratigraphy, archaeology, chronology, and past environments," *Journal of African Archaeology* 8(2), 185–218. The speculation that the Lebombo bone was used to measure the length of the menstrual cycle or the lunar cycle comes from P. Beaumont and R. Bednarik (2013), "Tracing the emergence of paleoart in sub-Saharan Africa," *Rock Art Research* 30(1), 33–54. This is a tantalizing idea, but given that the bone is broken at one end, we don't know if the person making the notches stopped at twenty-nine marks or made more notches that have been lost in the mists of time.

36. See M. Heun et al. (1997), "Site of Einkorn wheat domestication identified by DNA fingerprints," *Science* 278(5341), 1312–1314; S. Riehl et al. (2013), "Emergence of agriculture in the foothills of the Zagros Mountains of Iran," *Science* 341, 65–67; and D. Zohary (2012), *Domestication of plants in the Old World* (Oxford: Oxford University Press). On domestication more broadly, see G. Larson et al. (2014), "Current perspectives and the future of domestication studies," *Proceedings of the National Academy of Sciences* 111(17), 6139.

37. Consider these examples of agricultural systemizing: "*if* I have a ram and a ewe, *and* breed them, *then* I will have many sheep." Or, "*If* I have potato seeds, *and* plant them in early spring in moist warm soil, *then* I will have a supply of potatoes." Interestingly, agriculture was invented in at least eleven places independently. Note that there are lots of "ands" (or operations) you can perform on the input (the potato seeds) to maximize the output (getting lots of potatoes), including the depth at which you plant the seeds, the temperature of the seeds, when you harvest them, and so on. See "How to Grow Potatoes," Seed Savers Exchange, January 6, 2017, blog.seedsavers.org/blog/tips-for-growing-potatoes. These examples are adapted from Harari (2015), *Sapiens*.

38. Agriculture had many unintended consequences, as Harari points out in *Sapiens*. For example, being able to feed babies porridge and gruel enabled mothers to wean their babies earlier, and so they began to have a baby every year, because breastfeeding acts as a form of birth control. Weaning babies earlier also led to babies developing more infections, due to a weaker immune system. Child mortality increased to about one in three, and because agriculture often failed as well as succeeded, this led to malnutrition. In addition, communities grew. Consider, for example, how in 13,000 BC Jericho was home to only about 100 people, but by 8,000 BC Jericho had grown to 1,000 people. Agriculture also curbed the freedom of a nomadic lifestyle as people became tied to permanent homes. Finally, agriculture did not lead to working less—rather, humans now had to toil on the land and their quality of life became far worse. See S. LevYadun et al. (2000), "The cradle of agriculture," *Science* 288(5471), 1602–1603;

C. Larsen (2006), "The Agricultural Revolution as environmental catastrophe: Implications for health and lifestyle in the Holocene," *Quaternary International* 150(1), 12–20; and G. Barker (2006), *The Agricultural Revolution of prehistory: Why did foragers become farmers?* (Oxford: Oxford University Press).

39. A wheel is not just a circular object that can rotate, but also one that attaches to an axle bearing (a bearing being something that can support or bear something else). The question has often been raised as to why it took so long to invent the wheel. The common answer is that rotating a cylinder is not the hard part. The hard part is attaching a stable, stationary platform to it. The big breakthrough, according to anthropologist David Anthony, was the joining together of the wheel and the axle, so that the wheel could be used for transportation. He argues that several factors had to co-occur: carpentry using metal (copper) tools to chisel holes and axles, and a need for hauling heavy burdens over land. See D. Anthony (2007), *The horse, the wheel, and language* (Princeton, NJ: Princeton University Press).

40. An "armulet," a tablet of stone from 5,500 years ago that shows Old European Script, was found in Romania. The big breakthrough in Egypt 5,000 years ago enabled writing to be recorded on paper made from the papyrus plant— essentially writing on a leaf from the plant—or joined side by side to form a scroll that could be rolled up. Cave paintings are considered prehistorical because, although they use symbols, the symbols represent objects, such as a man or a horse, and are therefore not arbitrary linguistic symbols. See S. Houston (2004), *The first writing: Script invention as history and process* (Cambridge: Cambridge University Press); C. Walker (1987), *Cuneiform* (Berkeley: University of California Press); J. Allen (2013), *The ancient Egyptian language* (Cambridge: Cambridge University Press). Writing started as a "partial script," meaning that it had a limited set of symbols.

41. See S. Dehaene et al. (1998), "Abstract representations of numbers in the animal and human brain," *Trends in Neuroscience* 21(8), 355–361; and R. Wilder (1968), *Evolution of mathematical concepts: An elementary study* (New York: John Wiley & Sons).

42. See M. Bamshad et al. (2001), "Genetic evidence on the origins of Indian caste population," *Genome Research* 11, 904–1004.

43. Here's the algorithm: "*if* I take copper, *and* mix it with tin, *then* I get bronze, which is stronger than copper." A useful online resource about the Bronze Age is "The Bronze Age," SoftSchools.com, www.softschools.com/timelines/the _bronze_age_timeline/145/.

44. M. Roth (1995), *Law collections from Mesopotamia and Asia Minor* (Atlanta: Scholars Press).

45. By 3,300 BC, humans were developing systems for classifying herbs and concocting medicines. See J. Sumner (2000), *The natural history of medicinal plants* (Portland, OR: Timber Press). Early timepieces were also systemized: by 1300 BC, the Egyptians had built the sundial. See J. Bryner (2013), "Ancient Egyptian sundial discovered at Valley of the Kings," LiveScience, March 20, www .livescience.com/28057-ancient-egyptian-sundial-discovered.html. There is also

evidence of the early writing of oracles: by 1200 BC, Chinese oracles had been written of the form, "*if* a child is born, *and* it's on the date of x, *then* the child will be lucky." See K. Takashima (2012), "Literacy to the south and the east of Anyang in Shang China: Zhengzhou and Daxinzhuang," in F. Li and D. Branner, eds. *Writing and literacy in early China: Studies from the Columbia Early China Seminar*, (Seattle: University of Washington Press). By 205 BC, the Greeks had built the first computer (known as the Antikythera Mechanism), which could predict the movement of the planets and the sun and when eclipses would occur. See T. Freeth et al. (2006), "Decoding the ancient Greek astronomical calculator known as the Antikythera Mechanism," *Nature* 444(7119), 587–591. By the ninth century AD, Arabic numerals (from 0 to 9) had been invented by the Hindus, and Arabs had invaded India and brought back this numerical system to the Middle East and Europe. See L. Avrin (1991), *Scribes, script, and books: The book arts from antiquity to the Renaissance* (Chicago: American Library Association). And by the sixteenth century, formal science had been invented. See D. Wootton (2015), *The invention of science: A new history of the Scientific Revolution* (London: Penguin Random House UK).

46. See N. Goren-Inbar et al. (2004), "Evidence of hominin control of fire at Gesher Benot Ya'aqov, Israel," *Science* 304(5671), 725–727; W. Roebroeks and P. Villa (2001), "On the earliest evidence for habitual use of fire in Europe," *Proceedings of the National Academy of Science* 108(13), 5209–5214; and J. Gowlett (2016), "The discovery of fire by humans: A long and convoluted process," *Philosophical Transactions of the Royal Society: Series B, Biological Sciences* 371(1696), 20150164.

47. A. Gibbons (2007), "Food for thought: Did the first cooked meals help fuel the dramatic evolutionary expansion of the human brain?," *Science* 316(5831), 1558–1560; and L. Aiello and P. Wheeler (1995), "The expensive-tissue hypothesis: The brain and the digestive system in human and primate evolution," *Current Anthropology* 36(2), 199–221.

48. Six percent of modern Melanesian and Aboriginal Australian DNA is Denisovan DNA. Neanderthals left Africa 300,000 years ago. Denisovans split off from the Neanderthals about 400,000 years ago and settled in Europe and western Asia. In 2008, paleoanthropologists found a pinkie finger in a cave in Siberia belonging to a girl aged five to seven, from 40,000 years ago. Genetically, she was closely related to Neanderthal but distinct enough to warrant a new category, named Denisovan, after the cave where the pinkie was found. See M. Meyer et al. (2012), "A high-coverage genome sequence from an archaic Denisovan individual," *Science* 338(6104), 222–226; S. Pääbo (2015), "The diverse origins of the human gene pool," *Nature Reviews Genetics* 16(6), 313–314; S. Sankararaman (2014), "The genomic landscape of Neanderthal ancestry in present-day humans," *Nature* 507(7492), 354–357; D. Reich et al. (2010), "Genetic history of an archaic hominin group from Denisova Cave in Siberia," *Nature* 468, 1053–1060; M. Rasmussen et al. (2011), "An Aboriginal Australia genome reveals separate human dispersals into Asia," *Science* 334, 94–98; B. Vernot et al. (2016), "Excavating Neandertal and Denisovan DNA from the genomes of Melanesian

individuals," *Science*, March 17; and M. Kuhlwilm et al. (2016), "Ancient gene flow from early modern humans into eastern Neanderthals," *Nature* 530(7591), 429–433.

49. Bone harpoons have been found in the Democratic Republic of Congo from 90,000 years ago, and stone-lined hearths have been found in Zambia, from 100,000 years ago, the latter being strikingly absent from Neanderthal Europe. Carving a bone is likely to be specific to humans and would require complex tools to manufacture. See J. Yellen et al. (1995), "A Middle Stone Age worked bone industry from Katanda, Upper Semliki Valley, Zaire," *Science* 268, 553–556; and S. McBrearty and A. Brooks (2000), "The revolution that wasn't: A new interpretation of the origin of modern human behavior," *Journal of Human Evolution* 38, 453–563. In *The singing Neanderthals*, Mithen (2005) discusses how the first wooden spears were made 400,000 years ago. They were discovered in 1995 in Schoningen in southern Germany. See H. Thieme (1997), "Lower Paleolithic hunting spears from Germany," *Nature* 385, 807–810. Quite what the significance of these are remains unclear. The earliest evidence of "hafting"—where a stone point is attached to a wooden spear—dates back to 500,000 years ago. See J. Wilkins et al. (2012), "Evidence for early hafted hunting technology," *Science* 338, 942–946; and C. Baras (2012), "First stone-tipped spear thrown earlier than thought," *New Scientist*, November 15. This has been interpreted by different archaeologists as indicative that Neanderthals, or even *Homo heidelbergensis*, could make stone-tipped spears, not just modern *Homo sapiens*. Thomas Wynn and Frederick Coolidge (2012) in their book *How to think like a Neandertal* (Oxford: Oxford University Press) agree. We should keep open the possibility that Neanderthals could produce tools of the complexity of a wooden spear. A simpler flinthead spear could have been made with fewer steps (for example, making an ax and attaching it to a stick), which may account for why *Homo heidelbergensis* could have made these. The question is how these stone points were attached to a wooden spear.

50. See T. Higham et al. (2014), "The timing and spatiotemporal patterning of Neanderthal disappearance," *Nature* 512 (7514), 306–309; and C. Finlayson (2009), *The humans who went extinct: Why Neanderthals died out and we survived* (Oxford: Oxford University Press). The quote "the absence of evidence is not evidence of absence" has been attributed to my colleague Professor Sir Martin Rees, Trinity College, Cambridge.

51. There is some debate about whether Neanderthals made jewelry, as eagle talons were discovered dating back to 120,000 years ago. The debate is about whether this was actually jewelry. See M. Gannon (2015), "Neanderthals wore eagle talons as jewelry 130,000 years ago," LiveScience, March 11, www.live science.com/50114-neanderthals-wore-eagle-talon-jewelry.html.

52. See S. Baron-Cohen (1999), "The evolution of a theory of mind," in M. Corbalis and S. Lea, eds., *The descent of mind: Psychological perspectives on hominid evolution* (Oxford: Oxford University Press).

53. On mindreading and autism, see S. Baron-Cohen (1995), *Mindblindness* (Cambridge, MA: MIT Press). On systemizing and autism, see S. Baron-Cohen et

al. (2003), "The Systemizing Quotient: An investigation of adults with Asperger syndrome or high-functioning autism, and normal sex differences," *Philosophical Transactions of the Royal Society: Series B*, 358, 361–374.

CHAPTER 6:
SYSTEM-BLINDNESS: WHY MONKEYS DON'T SKATEBOARD

1. The earliest stone tools discovered to date go back to 3.3 million years ago. Stone tools are usefully divided into mode 1 stone choppers, which date to 3.3 to 2.5 million years ago and are crude and asymmetrical; mode 2 hand axes, from 2 million years ago; and mode 3 Mousterian tools from 0.4 million years ago, manufactured by Neanderthals. For mode 1, see S. Harmand et al. (2015), "3.3 million year old stone tools from Lomekwi 3, West Turkana, Kenya," Nature 521(7552), 310–315; and S. Semaw et al. (1997), "2.5 million year old stone tools from Gona, Ethiopia," Nature 385(6614), 333–336. For modes 2 and 3, see R. Klein (2009), The human career: Human biological and cultural origins, 3rd ed. (Chicago: University of Chicago Press). For a good review of human tool use, see C. Stringer (2012), "What makes a modern human?," Nature 485(7396), 33–35. See S. de Beaune et al., eds. (2009), *Cognitive archaeology and human evolution* (Cambridge: Cambridge University Press); and S. Boinski et al. (2008), "Substrate and tool use by Brown Capuchins in Suriname: Ecological contexts and cognitive bases," *American Anthropologist* 102(4), 741–761. For a useful online resource on animal tool use, see C. Choi (2011), "Creative creatures: 10 animals that use tools," LiveScience, November 3, www.livescience .com/16856-animals-tools-octopus-primates.html.

2. For crows cracking nuts, see David Attenborough, "Wild crows inhabiting the city use it to their advantage," BBC Wildlife, www.youtube.com/watch?v=BG PGKnpq3e0. See also A. Taylor et al. (2011), "New Caledonian crows learn the functional properties of novel tool types," *PLoS ONE*, December 4; and A. Auersberg et al. (2014), "Social transmission of tool use and tool manufacture in Goffin cockatoos (*Cacatua goffini*)," *Proceedings of the Royal Society of London: Series B* 281, 20140972.

3. See J. Fisher and R. Hinde (1949), "The opening of milk bottles by birds," *British Birds* 42, 347–357.

4. On animal learning, see Skinner (1938), *Behavior of organisms*.

5. On whether animals can innovate, see K. Laland (2017), *Darwin's unfinished symphony: How culture made the human mind* (Princeton, NJ: Princeton University Press), 100; and C. van Schaik et al. (1996), "Manufacture and use of tools in wild Sumatran orangutans," *Naturwissenschaften* 83, 186–188. For an interesting discussion, see S. Bhanoo (2014), "Chimpanzees' table manners vary by group," *New York Times*, May 12.

6. See S. J. Allen et al. (2011), "Why do Indo-Pacific bottlenose dolphins (*Tursiops sp.*) carry conch shells (*Turbinella sp.*) in Shark Bay, Western Australia?," *Marine Mammal Science* 27(2), 449–454; J. Mann et al. (2008), "Why do

dolphins carry sponges?.," *PLoS ONE* 3(12), e3868; and V. Gill (2014), "Cocka-toos teach tool-making tricks," BBC News, September 3, www.bbc.co.uk/news /science-environment-28990335.

7. See M. Haslam et al. (2019), "Wild sea otter mussel pounding leaves archae-ological traces," *Scientific Reports* 9, 4417.

8. See T. Breuer et al. (2005), "First observation of tool use in wild gorillas," *PLoS Biology* 3(11), e380.

9. See K. Wantanabe et al. (2007), "Long-tailed macaques use human hair as dental floss," *American Journal of Primatology* 69(8), 940–944.

10. See J. Wimpenny et al. (2009), "Cognitive processes associated with sequential tool use in New Caledonian crows," *PLoS ONE* 4(8), e6471.

11. See J. Plotnik et al. (2011), "Elephants know when they need a helping trunk in a cooperative task," *Proceedings of the American Academy of Science* 108(12), 5116–5121; and B. Hart et al. (2001), "Cognitive behaviour in Asian elephants: Use and modification of branches for fly switching," *Animal Behaviour* 62(5), 839–847.

12. See J. Finn et al. (2009), "Defensive tool use in a coconut-carrying octo-pus," *Current Biology* 19(23), R1069–R1070.

13. See A. Rutherford (2018), *The book of humans* (London: Weidenfeld and Nicholson); and M. Greshko (2018), "Why these birds carry flames in their beaks," *National Geographic*, January 8.

14. See D. Hanus and J. Call (2008), "Chimpanzees infer the location of a reward based on the effect of its weight," *Current Biology* 18, R370–R372.

15. I found a video of chimps ice-skating, but it's clearly the result of animal training—chimps don't surf, skate, or skateboard in the wild. See "Chimps on ice," YouTube, August 12, 2008, www.youtube.com/watch?v=pOj_QoSH6is. For a video of a crow snowboarding, see "Crowboarding: Russian roof-surfin' bird caught on tape," YouTube, www.youtube.com/watch?v=3dWw9GLcOeA. A snow-boarding crow can easily be interpreted as the result of reward learning, not evidence of invention. Finally, there's a video of snow macaques making snowballs and rolling them down slopes, but it looks more like social learning—the rest of the peer group are doing the same thing—than like clear evidence that they understand causality. See L. Young (2016), "Watch this adorable baby macaque roll a snowball down a hill," Atlas Obscura, December 16, www.atlasobscura .com/articles/watch-this-adorable-baby-macaque-roll-a-snowball-down-a-hill.

16. I'm coining the new term "system-blindness," which is akin to the term I coined in my book *Mindblindness*.

17. On preschoolers' causal understanding, see A. Gopnik and L. Schulz (2004), "Mechanisms of theory formation in young children," *Trends in Cognitive Sciences* 8(8), 371–377; A. Gopnik and L. Schulz (2007), *Causal learning: Psychol-ogy, philosophy, and computation* (New York: Oxford University Press); A. Taylor et al. (2014), "Of babies and birds: Complex tool behaviours are not sufficient for the evolution of the ability to create a novel causal intervention," *Proceed-ings of the Royal Society of London: Series B: Biological Sciences* 281(1787), 20140837; and K. M. Dewar and F. Xu (2010), "Induction, overhypothesis, and the origin

of abstract knowledge evidence from 9-month-old infants," *Psychological Science* 21(12), 1871–1877.

18. See F. Stewart et al. (2011), "Living archaeology: Artefacts of specific nest site fidelity in wild chimpanzees," *Journal of Human Evolution* 61(4), 388–395.

19. See D. Povinelli et al. (2000), *Folk physics for apes: The chimpanzee's theory of how the world works* (Oxford: Oxford University Press); and D. Penn and D. Povinelli (2007), "Causal cognition in human and non-human animals: A comparative, critical review," *Annual Review of Psychology* 58, 97–118. And on attempts to teach apes to use tools, see N. Toth et al. (1993), "Pan the tool-maker: Investigations into the stone tool making and tool-using capabilities of a bonobo (*Pan paniscus*)," *Journal of Archaeological Science* 20(1), 81–91.

20. Interestingly, the chimps eventually learned to use the rake that was the right way up, but it took them at least twenty-five trials. This suggests that they were using a different process—associative learning—to get the food, unlike even a young human child, who immediately grasps the causal significance of the rake that is the right way up. See D. Povinelli and S. Dunphy-Lelii (2001), "Do chimpanzees seek explanations? Preliminary comparative investigations," *Canadian Journal of Experimental Psychology* 55(2), 187–195.

21. See A. Bania et al. (2009), "Constructive and deconstructive tool modification by chimpanzees (*Pan Troglodytes*)," *Animal Cognition* 12, 85–95; I. Davidson and W. McGrew (2005), "Stone tools and the uniqueness of human culture," *Journal of the Royal Anthropological Institute* 11(4), 793–817.

22. Josep Call and his colleagues present some evidence that may challenge Povinelli's conclusions. See Hanus and Call (2008), "Chimpanzees infer the location of a reward"; and C. Volter et al. (2016), "Great apes and children infer causal relations from patterns of variation and covariation," *Cognition* 155, 30–43.

23. On causality, see M. Lombard and P. Gärdenfors (2017), "Tracking the evolution of causal cognition in humans," *Journal of Anthropological Sciences* 95, 1–16. Lombard and Gärdensfors call this "causal grammar," but I simply call it systemizing, or *if-and-then* pattern seeking.

24. On the bow-and-arrow, see Brown et al. (2011), "An early and enduring advanced technology"; and M. Lombard and M. Haidle (2012), "Thinking a bow-and-arrow set: Cognitive implications of Middle Stone Age bow and stone-tipped arrow technology," *Cambridge Archaeological Journal* 22, 237–264.

25. Although some chimps have been observed throwing stones at trees, it has been suggested that they may be doing so simply to attract females, without understanding cause-and-effect. See H. Kuhl et al. (2016), "Chimpanzee accumulative stone throwing," *Scientific Report* 6, article 22219. In 1975, P. J. Darlington of Harvard University's Museum of Comparative Zoology claimed that throwing with precision or accuracy is a uniquely human trait. Darlington described a study in which wild chimpanzees threw forty-four objects but only successfully struck their target five times, and then only when they were within 2 meters (6.6 feet) of it. "Other primates do throw sticks and stones, but only awkwardly. . . . Compare this with human throwing. A skilful man has a good chance to break the skull of another man with one stone at 30m (100ft)," he added.

Cited in J. Goldman (2014), "Can humans throw better than animals?," BBC, February 25. Humans alone can reason: "*if* my dart is gripped between my finger and thumb, *and* I close one eye so the tip of the dart is in line with the bull's-eye, *and* I throw the dart along an arc and release, *then* it will hit the bull's-eye." The idea that apes may not throw because of their hand morphology is discussed in J. Wood et al. (2007), "The uniquely human capacity to throw evolved from a non-throwing primate: An evolutionary dissociation between action and perception," *Biology Letters* 3, 360–364. One chimpanzee was studied that seemed to be stockpiling stones and hiding them to throw at visitors in a zoo, but this seems to have been more of a dominance display and the *interpretation* of why the chimpanzee was stockpiling and hiding the stones may be just that. See M. Osvath and E. Karvonen (2012), "Spontaneous innovation for future deception in a male chimpanzee," *PLoS ONE* 7(5), e36782.

26. At the risk of repetition, these examples of animal tool use could just be explained by associative learning and do not imply that the animal understands if-and-then patterns, particularly where the "and" is a causal operation. As a reminder of what associative learning is, imagine an animal learns that ringing a doorbell (A) is associated with the door opening (B). The animal could learn this through pairing A with B, without any necessary understanding of causality, or indeed any understanding that there is a system. In contrast, when you or I ring the doorbell, we do this using *if-and-then* reasoning. For example, we might think: "*if* the door is closed, *and* I ring the bell, *and* there's someone on the other side who heard it and wants to let me in, *then* the door will open." The *and* is the key causal "operation" between the *if* (A) and the *then* (B). That little word *and* represents a world of difference. It represents a causal operation. That's why, if we ring and the door doesn't open, we start thinking of *explanations*: "perhaps there's no one home," or "perhaps they haven't heard the bell," or "perhaps the bell isn't working." In this example, we are using our Empathy Circuit (to imagine the thoughts and perceptions of the person on the other side of the door) as well as our Systemizing Mechanism (to figure out what causes the door to open). There's no evidence in animal behavior that non-human animals seek explanations or are curious about identifying them. Associative learning is also referred to as Edward Thorndike's "Law of Effect": an animal will try out different behaviors and retains only those whose effects lead to the outcome they wanted. See E. Thorndike (1898), "Animal intelligence: An experimental study of the associative processes in animals," *Psychological Monographs* 8.

27. On the earliest sharp-tipped arrows, see M. Lahr et al. (2016), "Inter-group violence among early Holocene hunter-gatherers of West Turkana, Kenya," *Nature* 529(7586), 394–398; and Brown et al. (2011), "An early and enduring advanced technology."

28. See S. Carounanidy, "Sophisticated time-awareness: The human spark?," What Makes Us Human, whatmakesushumans.com/category/human-uniqueness /page/5/. Nine is the minimum number of steps to make a bow-and-arrow, and this depends on the design. For a fifteen-step process, the chance of a tool arising by chance alone is over one in a billion.

CHAPTER 7: THE BATTLE OF THE GIANTS

1. See F. Max Müller (1861), *Lectures on the science of language delivered at the Royal Institution of Great Britain in April, May, and June, 1861* (London: Longman, Green, Longman & Roberts); and Darwin, *The descent of man.*

2. On language driving invention, see S. Mithen (1996), *The prehistory of the mind: The search for the origins of art, religion, and science* (London: Thames and Hudson Ltd.).

3. On the evolution of language, see R. Botha and C. Knight (2009), *The cradle of language* (Oxford and New York: Oxford University Press); C. Perreault and S. Mathew (2012), "Dating the origin of language using phonemic diversity," *PLoS ONE* 7(4), e35289; M. Dunn et al. (2011), "Evolved structure of language shows lineage-specific trends in word-order universals," *Nature* 473(7345, May), 79–82; Q. Atkinson (2011), "Phonemic diversity supports a serial founder effect model of language expansion from Africa," *Science* 332(6027), 346–349; R. Berwick and N. Chomsky (2016), *Why only us: Language and evolution* (Cambridge, MA: MIT Press); and R. Burling (2007), *The talking ape* (Oxford: Oxford University Press).

4. The hyoid bone is markedly different in modern chimpanzees and in *Australopithicus.* See B. Arensburg et al. (1989), "A Middle Paleolithic human hyoid bone," *Nature* 338(6218), 758–760; L. Capasso et al. (2008), "A *Homo erectus* hyoid bone: Possible implications for the origin of the human capability for speech," *Collegium Antropologicum* 32(4), 1007–1011; D. Dediu and S. Levinson (2013), "On the antiquity of language: The reinterpretation of *Neanderthal* linguistic capacities and its consequences," *Frontiers in Psychology* 4; D. Frayer (2017), "Talking hyoids and talking *Neanderthals*," in E. Delson and E. Sargis, eds., *Vertebrate paleobiology and paleoanthropology series* (Springer); and D. Frayer and C. Nicolay (2000), "Fossil evidence for the origin of speech sounds," in N. Wallin et al., eds., *The origins of music* (Cambridge, MA: MIT Press). Note that the hyoid bone is not the only relevant part of the speech apparatus. So is increased voluntary control of the respiratory muscles, which is present in Neanderthals but not *Homo erectus.*

5. See J. Riley et al. (2005), "The flight paths of honeybees recruited by the waggle dance," *Nature* 435(7039), 205–207; and J. Nieh (2004), "Recruitment communication in stingless bees (Hymenoptera, Apidae, Meliponini)," *Apidologie* 35(2), 159–182. On birds singing at dawn, see J. Hutchinson (2002), "Two explanations of the dawn chorus compared: How monotonically changing light levels favour a short break from singing," *Animal Behaviour* 64, 527–539.

6. Communication in non-human animals is likely to be different to what we call communication in humans, since in humans communication involves "reference," which requires a theory of mind, part of the Empathy Circuit in the human brain. When a human speaker uses the word "bowl" and points to this object, the speaker intends their audience to *recognize* their *intention* for the sound "bowl" to refer to a bowl. In contrast, when a vervet makes a different alarm call in the presence of a tiger, this serves as a useful early-warning system letting other monkeys who hear the alarm know what to do, but this is no evidence that the "speaker"

is referring to anything. Pika and Mitani (2006) suggest that chimps can point to the part of their body they want another chimp to scratch during grooming, as if to say "Scratch here," and that this is understood as intentional communication and reference by the receiver. A more parsimonious explanation is that chimps learn that if they point to where they want to be scratched, they get scratched there, so reward learning can explain this without assuming they have a capacity for reference. If they could refer, we would see them referring widely, not just to where they want to be scratched. See S. Pika and J. Mitani (2006), "Referential gestural communication in wild chimpanzees (Pan troglodytes)," *Current Biology* 16(6), R191–R192; D. L. Cheney and R. M. Seyfarth (1990), *How monkeys see the world* (Chicago: University of Chicago Press); and H. Grice (1989), *Studies in the way of words* (Cambridge, MA: Harvard University Press).

7. See M. Hauser et al. (2002), "The faculty of language: What is it, who has it, and how did it evolve?," *Science* 298, 1569–1579. For a discussion on the definition of recursion, see "What is recursion?," Linguistics, linguistics.stack exchange.com/questions/3252/what-is-recursion; see also A. Vyshedskiy (2019), "Language evolution to revolution: The leap from rich-vocabulary non-recursive communication system to recursive language 70,000 years ago was associated with acquisition of a novel component of imagination, called prefrontal synthesis, enabled by a mutation that slowed down the prefrontal cortex maturation simultaneously in two or more children—the Romulus and Remus hypothesis," *Research Ideas and Outcomes* 5, e35846.

8. See I. Cross (2001), "Music, mind, and evolution," *Psychology of Music* 29, 95–102. Examples of recursion in music would include a verse (A), followed by a chorus (B), and returning to a verse (A) again: ABA. This ABA sequence may then become nested in a more complex structure, such as AA ABA AA. A different example of recursion in music might be within a verse, where you may hear a sequence of notes (A), then a second sequence (B), and then the first reappears (A), perhaps with a subtle variation.

9. On the relationship between language and music: Neuropsychology often uses single case-studies to make inferences about whether two functions in the brain are independent of each other. The following are illustrative:

Alexander Luria, the Russian neuropsychologist, reported a case of a patient, Shebalin, who was a distinguished musician who lost his language after a stroke but not his musical ability. During his aphasia, he continued to compose. This shows that music and language are represented independently in the brain.

Another one of Luria's neurological cases, "NS," had a stroke and lost his capacity to understand simple phrases but could still recognize melodies and sing them. This confirms the independence of language and music in the brain.

And there are musically talented individuals who never develop much language, often called autistic savants, again demonstrating the independence of music and language in the brain. See A. Luria et al. (1965), "Aphasia in a composer," *Journal of Neurological Science* 2, 288–292.

Neuropsychologist Isabelle Peretz reports that her patient "HJ" had a stroke that left him with severe amusia (an inability to recognize music) even though his language was intact. Again, this confirms the independence of language and music in the brain. See I. Peretz and K. L. Hyde (2003), "What is specific to music processing? Insights from congenital amusia," *Trends in Cognitive Sciences* 7(8), 362–367; and M. Mendez (2001), "Generalized auditory agnosia with spared music recognition in a left hander: Analysis of a case with a right temporal stroke," *Cortex* 37, 139–150.

People who lose an aspect of their musical ability (amusia) as a result of a stroke typically lose pitch perception but not the ability to perceive or invent rhythm. This likely reflects the dependence of rhythm on the Systemizing Mechanism.

Mithen discusses the case of Monica, an intelligent, well-educated woman who had amusia: she responded to music as if it were noise and had never been able to sing or dance. Monica's amusia was primarily in pitch perception, not in rhythm perception, which is what one would predict from the universality of systemizing in humans. A series of other cases of congenital amusia have confirmed the independence of pitch perception and rhythm perception (systemizing).

See Mithen, *The singing Neanderthals*; S. Wilson and J. Pressing (1999), "Neuropsychological assessment and modelling of musical deficits," *Music Medicine* 3, 47–74; I. Peretz et al. (2002), "Congenital amusia: A disorder of fine grained pitch perception," *Neuron* 33, 185–191; J. Ayotte et al. (2002), "Congenital amusia," *Brain* 125, 238–251; and M. Thaut et al. (2014), "Human brain basis of musical rhythm perception: Common and distinct neural substrates for meter, tempo, and pattern," *Brain Sciences* 4, 428–452.

10. I have to briefly deal with the claim by linguist Steven Pinker that music is "auditory cheesecake." See S. Pinker (1994), *The language instinct: How the mind creates language* (New York: William Morrow). What he means is that our love of music is simply a by-product of something more basic, such as hearing or spoken language, both of which were adaptive. When something is a by-product of evolution, it is called an "exaptation" rather than an adaptation. Thus, according to Pinker, our love of fat and sugars was an adaptation, but our love of cheesecake is an exaptation. He argues: "What benefit could there be to diverting time and energy to making plinking noises?...As far as biological cause and effect are concerned, music is useless....It could vanish from our species and the rest of our lifestyle would be virtually unchanged." My view is different: I argue that music is an outward indicator that, as a species, we are *if-and-then* pattern seekers. Music could vanish only if our species lost its Systemizing Mechanism, which would mean we would also have lost the capacity for invention.

11. Mithen argues that "the human mind evolved to enjoy melody and rhythm, which were critical features of communication before becoming usurped by language" (*The Singing Neanderthals*). I think this is back-to-front: as soon as we could enjoy melody and rhythm, we had the ingredients for invention, technology, and syntax, because we could take things apart (segmentation and compositionality). Mithen speculates that Neanderthals could enjoy melody and rhythm, but I'm not convinced, because if they could enjoy these, then they should have been able to systemize and, thereby, invent.

12. See M. Zentner and T. Eerola (2010), "Rhythmic engagement with music in infancy," *Proceedings of the National Academy of Sciences* 107(13), 5768–5773; I. Winkler et al. (2009), "Newborn infants detect the beat in music," *Proceedings of the National Academy of Sciences* 106(7), 2468–2471. On non-human animals' rhythm and music perception, see M. Hauser and J. McDermott (2003), "The evolution of the music faculty: A comparative perspective," *Nature Neuroscience* 6, 663–668. Although some research found that female parakeets prefer to spend more time in a rhythmic booth than an arrhythmic booth, this small sample needs to be replicated and doesn't prove that they get the beat. See M. Hoeschele and D. Bowling (2016), "Sex differences in rhythmic preferences in the Budgerigar (*Melopsittacus undulatus*): A comparative study with humans," *Frontiers in Psychology* 7, 1543.

13. Some argue that the unique features of language are segmentation and compositionality, which, according to philosopher Peter Carruthers, are achieved through syntax. See A. Wray (1998), "Protolanguage as a holistic system for social interaction," *Language and Communication* 18, 47–67; P. Carruthers (2002), "The cognitive function of language," *Brain and Behavioural Sciences* 25, 657–726.

14. See L. Selfe (1977), *Nadia: A case of extraordinary drawing ability in an autistic child* (Cambridge, MA: Academic Press); L. Selfe (2011), *Nadia revisited: A longitudinal study of an autistic savant* (London: Psychology Press); and S. Wiltshire (1989), *Cities* (London: J. M. Dent and Sons Ltd.).

15. The language system in the human brain also of course draws on many other neural processes, but here is not the place to digress further into the complexity of language itself. For an excellent overview of language, see Pinker, *The language instinct.*

16. Vyshedskiy argues that the function of the lateral prefrontal cortex is to integrate two concepts. Damage to the lateral prefrontal cortex results in prefrontal aphasia, also called frontal dynamic aphasia. He says that these patients struggle with questions like "If a cat ate a dog, who is alive?" or "Imagine a blue cup inside a yellow cup: which cup is on top?" Fuster says that these patients can keep a conversation going and understand single words or short sentences but struggle to "propositionalize." Note that the lateral prefrontal cortex is also involved in planning, behavioral inhibition, set-shifting, and decision-making. See Vyshedskiy, "Language evolution to revolution"; M. Watanable (2009), "Role of the primate lateral prefrontal cortex in integrating decision-making and motivational information," in J. C. Dreher and L. Tremblay, eds., *Handbook of reward and decision making* (Burlington, MA: Academic Press); A. Friederici (2011), "The brain basis of language processing: From structure to function," *Physiological Review* 91, 1357–1392; J. Fuster (2008), "Human neuropsychology," in *The prefrontal cortex* 4th ed. (Cambridge, MA: Academic Press); A. Luria (1970), *Traumatic aphasia* (Berlin and Boston: De Gruyter Mouton).

17. Vyshedskiy argues that prefrontal synthesis is also needed to understand the phrase "A snake on the boulder to the left of the tall tree that is behind the hill," because the listener has to combine four objects (snake, boulder, tree, hill) into a novel scene. Such phrases clearly require recursion, in the sense of nesting.

However, understanding this can also be achieved by *if-and-then* reasoning: *if* the snake is on the boulder, *and* the boulder is to the left of the tall tree, *and* the tree is behind the hill, *then* the snake is behind the hill, etc. He also argues that prefrontal synthesis is close to what Chomsky calls "merge": the ability to take any two "syntactic" objects to create a new one (for example, "house-boat"). See N. Chomsky (2008), "On phrases," in R. Freidin et al., eds., *Foundational issues in linguistic theory: Essays in honor of Jean-Roger Vergnaud* (Cambridge, MA: MIT Press).

18. See A. Nowell (2010), "Defining behavioral modernity in the context of Neandertal and anatomically modern human populations," *Annual Review of Anthropology* 39, 437–452; and I. Tattersall (2006), "How we came to be human," *Scientific American,* June 1. I am indebted to the late Rick Cromer, who held evening seminars at the MRC Cognitive Development Unit at University College London, back in 1982–1983, discussing the question of how we are capable of symbol-thinking.

19. This logic also holds for any other sentence that is prefixed with a mental state. See A. M. Leslie (1987), "Pretense and representation: The origins of 'theory of mind,'" *Psychological Review* 94, 412–426. The special suspension of truth conditions that mental states confer on a proposition is called "referential opacity."

20. Y. Harari (2015), *Sapiens.*

21. There were some places on the planet that decided not to go into "lockdown," such as Sweden, but lockdown was effective across huge populations including India and China.

22. This theory is discussed in F. Wynn and T. Coolidge, *How to think like a Neandertal.* See C. Raby et al. (2007), "Planning for the future by Western scrub-jays," *Nature* 445, 919–921; and N. J. Mulcahy and J. Call (2006), "Apes save tools for future use," *Science* 312, 1038–1040.

23. C. Zimmer (2007), "Time in the animal mind," *New York Times,* April 3.

24. See L. Leakey et al. (2004), "A new species of genus Homo from Olduvai Gorge," *Nature* 4(202), 7–9; Klein, *The human career;* T. Feix et al. (2015), "Estimating thumb-index finger precision grip and manipulation potential in extant and fossil primates," *Journal of the Royal Society: Interface* 12(106), 20150176; A. Bardo et al. (2018), "The impact of hand proportions on tool grip abilities in humans, great apes, and fossil hominins: A biomechanical analysis using musculoskeletal simulation," *Journal of Human Evolution* 125, 106–121; and C. Kuzawa and J. Bragg (2012), "Plasticity in human life history strategy: Implications for contemporary human variation and the evolution of genus Homo," *Current Anthropology* 53(S6), S369–S382.

25. See S. McBrearty and A. Brooks (2000), "The revolution that wasn't: A new interpretation of the origin of modern human behavior," *Journal of Human Evolution* 39(5), 453–563; and J. Zilhao (2010), "Symbolic use of marine shells and mineral pigments by Iberian Neandertals," *Proceedings of the National Academy of Sciences* 107(3), 1023–1028. See J. Baio et al. (2018), "Prevalence of autism spectrum disorder among children aged 8 years—Autism and Developmental Disabilities Monitoring Network, 11 sites, United States, 2014," *MMWR Surveillance Summary* 67 (SS-6), 1–23.

CHAPTER 8: SEX IN THE VALLEY

1. See J. Baio et al. (2018), "Prevalence of autism spectrum disorder among children aged 8 years—Autism and Developmental Disabilities Monitoring Network, 11 sites, United States, 2014," *MMWR Surveillance Summary* 67 (SS-6), 1–23.

2. See S. Baron-Cohen et al. (1997), "Is there a link between engineering and autism?," *Autism* 1, 101–108; and R. Grove et al. (2015), "Exploring the quantitative nature of empathy, systemising, and autistic traits using factor mixture modelling," *British Journal of Psychiatry* 207, 400–406.

3. See S. Baron-Cohen and J. Hammer (1997), "Parents of children with Asperger syndrome: What is the cognitive phenotype?," *Journal of Cognitive Neuroscience*, 9, 548–554; G. Windham et al. (2009), "Autism spectrum disorders in relation to parental occupation in technical fields," *Autism Research* 2(4), 183–191.

4. Jim Simons became chair of Stony Brook University's math department at age thirty. He and Shiing-Shen Chern published the landmark Chern-Simons invariants, which have applications in quantum field theory, condensed-matter physics, and even string theory. He went on to win the American Mathematical Society's Oswald Veblen Prize, geometry's highest honor, for his research on area-minimizing surfaces. See A. Schaffer (2016), "The polymath philanthropist," *MIT Technology Review*, October 18, www.technologyreview.com/s/602561/the-polymath-philanthropist/.

5. Steve Shirley's original name was Stephanie, but she found that when she signed her letters "Steve" back in the 1960s, the contracts with her software company started pouring in. See S. Shirley and R. Asquith (2018), *Let IT go: The memoirs of Dame Stephanie Shirley* (Wilton, NH: Acorn Books).

6. This information was provided by Kurt Schöffer, CEO of Auticon, in a personal communication after he attended an annual meeting of these parents organized by the company UBS.

7. See L. Hawking (2015), "Dear Katie Hopkins. Please stop making life harder for disabled people," *Guardian*, April 30, www.theguardian.com/commentisfree/2015/apr/30/katie-hopkins-life-harder-disabled-people.

8. Elon Musk's ex-wife Justine Musk talks about their autistic son in a TEDx talk, embedded within Quora. Justine says their child was diagnosed with mild to moderate autism at age 4 but is now considered off the spectrum. See www.quora.com/Does-Elon-Musk-have-an-autistic-son-Which-one.

9. See B. Hughes (2003), "Understanding our gifted and complex minds: Intelligence, Asperger's syndrome, and learning disabilities at MIT," MIT Alumni Association newsletter. Note that in the United States the Institutional Review Board (IRB) is the ethics committee.

10. A clear example of where universities can set limits on research its academic staff undertake is the Chinese scientist who was sacked for cloning a human embryo. See P. Rana (2019), "How a Chinese scientist broke the rules to create the first gene edited babies," *Wall Street Journal*, May 10, www.wsj.com/articles/how-a-chinese-scientist-broke-the-rules-to-create-the-first-gene-edited-babies-11557506697.

11. Fiona Matthews, Rosa Hoekstra, Carol Brayne, and Carrie Allison made up the Eindhoven Study team. Each of the three Dutch cities has a population of over 200,000 people, and they were comparable in average income per household, percentage of children with special needs, and rates of mental health challenges. For comparison, we also asked about rates of two other disabilities: attention deficit hyperactivity disorder (ADHD) and dyspraxia (physical clumsiness). In the Eindhoven Study, all we had were school record data, so the parental occupations of the autistic children there remains an assumption that needs to be tested. But it seems a safe assumption, given the nature of Eindhoven. See M. Roelfsema et al. (2012), "Are autism spectrum conditions more prevalent in an information-technology region? A school-based study of three regions in the Netherlands," *Journal of Autism and Developmental Disorders* 42, 734–739.

12. Rates of autism in Silicon Valley still need to be studied, but because health services are not centralized in the United States, with each family using a different health insurance provider, it is not easy to gather information there.

13. See J. Erlandsson and K. Johannesson (1994), "Sexual selection on female size in a marine snail, *Littorina littorea* (L.)," *Journal of Experimental Marine Biology and Ecology* 181, 145–157; A. Fargevieille et al. (2017), "Assortative mating by colored ornaments in blue tits: Space and time matter," *Ecology and Evolution* 7(7), 2069–2078; G. Stulp et al. (2017), "Assortative mating for human height: A meta-analysis," *American Journal of Human Biology* 29(1, January–February), e22917; K. Han, N. C. Weed, and J. N. Butcher (2003), "Butcher dyadic agreement on the MMPI-2," *Personality and Individual Differences* 35, 603–615; and J. Glickson and H. Golan (2001), "Personality, cognitive style, and assortative mating," *Personality and Individual Differences* 30, 1109–1209.

14. See S. Baron-Cohen (2006), "Two new theories of autism: Hyper-systemizing and assortative mating," *Archives of Diseases in Childhood* 91, 2–5; S. Connolly et al. (2019), "Evidence of assortative mating in autism spectrum disorder," *Biological Psychiatry* 86(4), 286–293; A. E. Nordsletten et al. (2016), "Patterns of nonrandom mating within and across 11 major psychiatric disorders," *JAMA Psychiatry* 73(4), 354–361; J. Wouter et al. (2016), "Exploring boundaries for the genetic consequences of assortative mating for psychiatric traits," *JAMA Psychiatry* 73(11), 1189–1195.

15. Assortative mating could also explain why parents of autistic children tend to be older than average—it has just taken them longer to find someone who will marry them. See www.spectrumnews.org/news/link-parental-age-autism -explained. For example, men over fifty-five are four times as likely to have an autistic child as men under thirty. See C. Hultman (2011), "Advancing paternal age and risk of autism: New evidence from a population-based study and a meta-analysis of epidemiological studies," *Molecular Psychiatry* 16(12), 1203–1212. Of course, this doesn't rule out an age-related effect on the germline.

16. Note that all three of these explanations for assortative mating in autism could be correct and be happening simultaneously: hyper-systemizers of opposite sexes could be moving to the same location *and* find themselves in a smaller

pool of possible mates because of social skills *and* be attracted to a like-minded person who loves details and if-and-then pattern seeking, just as they do.

17. See S. Baron-Cohen and J. Hammer (1997), "Parents of children with Asperger syndrome: What is the cognitive phenotype?," *Journal of Cognitive Neuroscience* 9, 548–554; T. Jolliffe and S. Baron-Cohen (1997), "Are people with autism or Asperger syndrome faster than normal on the Embedded Figures Task?," *Journal of Child Psychology and Psychiatry* 38, 527–534.

CHAPTER 9:
NURTURING THE INVENTORS OF THE FUTURE

1. See D. L. Christensen et al. (2019), "Prevalence and characteristics of autism spectrum disorder among 4-year-old children—Early Autism and Developmental Disabilities Monitoring Network, seven sites, United States, 2010–2014," *MMWR Surveillance Summary* 68(SS-2), 1–19; and Baio et al., "Prevalence of autism spectrum disorder among children aged 8 years."

2. See S. Baron-Cohen (2019), "The concept of neurodiversity is dividing the autism community," *Scientific American,* April 30; D. Muzikar (2018), "Neurodiversity: A person, a perspective, a movement?," The Art of Autism, September 11, the-art-of-autism.com/neurodiverse-a-person-a-perspective-a-movement/; and S. Baron-Cohen (2017), "Neurodiversity: A revolutionary concept for autism and psychiatry," *Journal of Child Psychology and Psychiatry* 58, 744–747. At Cambridge, I taught a course called Abnormal Psychology from the mid-1990s; I renamed it Atypical Psychology in 2010, because the word "abnormal" had become dated and was associated with pathologizing those who are different or atypical. Even today there is still a scientific journal called the *Journal of Abnormal Psychology,* so this idea hasn't entirely gone away.

3. See *Creative differences: A handbook for embracing neurodiversity in the creative industries* (Universal Music, January 2020). This handbook also helps employers implement "reasonable adjustments" in the workplace to promote neurodiversity.

4. There is no evidence that Einstein ever said this. But see Quote Investigator, quoteinvestigator.com/2013/04/06/fish-climb.

5. See Baron-Cohen, "Neurodiversity: A revolutionary concept for autism and psychiatry"; and Baron-Cohen, "The concept of neurodiversity is dividing the autism community."

6. On the history of the concept of neurodiversity, see H. Bloom (1997), "Neurodiversity: On the neurological advantages of Geekdom," *Atlantic,* September; and J. Singer (2016), *Neurodiversity: The birth of an idea* (Amazon.com Services LLC).

7. Note that more recent studies suggest that the unemployment rate may still be unacceptably high but is more like 60 percent, not as high as 85 percent. See B. Reid (2006), *Moving on up? Negotiating the transition to adulthood for young people with autism* (London: National Autistic Society); J. Barnard et al. (2001), *Ignored and ineligible? The reality for families with autistic spectrum disorders* (London: National Autistic Society); and Griffiths et al., "The Vulnerability Experiences Quotient (VEQ)."

8. In addition to the companies discussed here that welcome autistic people—Specialisterne (specialisterne.com), Auticon (auticon.co.uk), and SAP (www .sap.com/corporate/en/company/diversity/differently-abled.html)—other companies include Deloitte, Universal Music, Microsoft, HP Enterprise, IBM, the accounting firms Ernst & Young and PricewaterhouseCoopers, Ford, Freddie Mac, DXC Technology, and the UK signals intelligence agency GCHQ. For a list of autism-friendly companies, see Disability:IN, "Autism @ Work Employer Roundtable," usbln.org/what-we-do/autism-employer-roundtable.

9. See Neurotribes and Steve Silberman's TEDtalk, "The forgotten history of autism," March 2015, www.ted.com/talks/steve_silberman_the_forgotten _history_of_autism?language=en.

10. Quoted in J. Chu (2017), "Why SAP wants to train and hire nearly 700 adults with autism," *Inc.*, www.inc.com/jeff-chu/sap-autism-india.html.

11. See R. Austin and G. Pisano (2017), "Neurodiversity as a competitive advantage," *Harvard Business Review* (May–June).

12. See C. Hall (2017), "Neurodiverse like me," Medium, April 5, medium .com/sap-tv/robots-and-people-autism-at-work-c7fc40e4d39a.

13. See C. Best et al. (2015), "The relationship between subthreshold autistic traits, ambiguous figure perception, and divergent thinking," *Journal of Autism and Developmental Disorders* 5(12), 4064–4067.

14. See Y. Lappin (2018), "The IDF's Unit 9900: 'Seeing' their service come to fruition," Jewish News Syndicate, May 4, www.jns.org/the-idfs-unit-9900-seeing -their-service-come-to-fruition/; and S. Rubin (2016), "The Israeli army unit that recruits teens with autism," *Atlantic*, January 6. On autistic people checking security, see C. Gonzales et al. (2013), "Practice makes improvement: How adults with autism out-perform others in a naturalistic visual search task," *Journal of Autism and Developmental Disorders* 43(10), 2259–2268.

15. James Neely quoted in J. Harris (2017), "How do you solve the trickiest problems in the workplace? Employ more autistic people," *Guardian*, October 9.

16. I use the term "learning disability" rather than "learning difficulty" to refer to below-average IQ, in line with current practice. The guidance is that the term "learning difficulty" should refer to a specific disability, such as ADHD, that does not affect global IQ. See NHS, "Overview: Learning disabilities," www.nhs .uk/conditions/learning-disabilities/; and "Learning Difficulties," mencap, www .mencap.org.uk/learning-disability-explained/learning-difficulties.

17. See T. Clements (2017), "The problem with the Neurodiversity Movement," *Quillette*, October 15. Jonathan Mitchell is an autistic man who asserts, "We don't want no neurodiversity," on his blog, Autism's Gadfly, autismgadfly .blogspot.com. Some are anti-neurodiversity because they believe that autism is caused by vaccines, a position that has been repeatedly refuted by scientific studies. See K. Knight (2018), "I'm autistic—Don't let anti-vaxxers bring the culture of fear," *Guardian*, August 23, www.theguardian.com/commentisfree/2018/aug /23/autistic-anti-vaxxers-fear-neurodiversity-far-right.

18. On rates of gastrointestinal pain in autistic people, see V. Chaidez et al. (2014), "Gastrointestinal problems in children with autism, developmental

delays, or typical development," *Journal of Autism and Developmental Disorders* 44(5), 117–127. On rates of epilepsy in autistic people, see J. Perrin et al. (2016), "Healthcare for children and youth with autism and other neurodevelopmental disorders," *Pediatrics* 137(suppl. 2).

19. See Simons Foundation, "SFARI gene," www.sfari.org/resource/sfari -gene/; Warrier and Baron-Cohen, "The genetics of autism"; and Huguet et al., "The genetics of autism spectrum disorders."

20. On autism and prematurity, see N. Padilla et al. (2015), "Poor brain growth in extremely preterm neonates long before the onset of autism spectrum disorder symptoms," *Cerebral Cortex* 27, 1245–1252. On autism and birth complications, see S. Jacobsen et al. (2017), "Association of perinatal risk factors with autism spectrum disorder," *American Journal of Perinatology* 34(3), 295–304.

21. See G. Owens et al. (2008), "LEGO® therapy and the social use of language programme: An evaluation of two social-skills interventions for children with high functioning autism and Asperger syndrome," *Journal of Autism and Developmental Disorders* 38, 1944–1957; and D. Legoff et al. (2014), *Lego Therapy: How to build social competence for children with autism and related conditions* (London: Jessica Kingsley Ltd.).

22. See O. Golan et al. (2010), "Enhancing emotion recognition in children with autism spectrum conditions: An intervention using animated vehicles with real emotional faces," *Journal of Autism and Developmental Disorders* 40, 269–279; and Cambridge Autism Learning, "Training and resources for parents and carers of autistic children," www.cambridgeautismlearning.com. An example of systemizing in this animation is, *if* Charlie-the-train's face is neutral, *and* he gets stuck halfway up the hill, *then* his face will change to angry.

23. See O. Golan and S. Baron-Cohen (2006), "Systemizing empathy: Teaching adults with Asperger syndrome or high-functioning autism to recognize complex emotions using interactive multimedia," *Development and Psychopathology* 18, 591–617; and Cambridge Autism Learning, www.cambridgeautismlearning .com.

24. See E. Glettner (2013), "Skateboarding is therapeutic for autistic children," *Huffington Post*, February 6. An example of systemizing skateboarding might be, "*if* I lean back, *and* push down, *then* the nose of my skateboard will rise off the ground."

25. On Disney movies and autism, see R. Suskind (2014), *Life animated: A story of side-kicks, heroes, and autism* (Los Angeles: Kingswell). This is one example of what Suskind calls "affinity therapy": find what every autistic kid is passionate about, which usually involves hyper-systemizing, and use that as the way to connect to them.

26. On the rate of obsessive compulsive disorder in autism, see M. C. Lai et al. (2014), "Autism," *Lancet*, 383, 896–910; T. Cadman et al. (2015), "Obsessive-compulsive disorder in adults with high-functioning autism spectrum disorder: What does self-report with the OCI-R tell us?," *Autism Research* 8, 477–485; F. van Steensel et al. (2011), "Anxiety disorders in children and adolescents with autistic spectrum disorders: A meta-analysis," *Clinical Child and Family Psychology Review*

14(3), 302–317; and V. Postorino et al. (2017), "Anxiety disorders and obsessive-compulsive disorder in individuals with autism spectrum disorder," *Current Psychiatry Reports* 19(12), 92.

27. Hyper-systemizers (Extreme Type S) are just 2.5 percent of kids, and some kids who are in the borderlands (Type S) might wish to also join this educational stream.

28. On Greta Thunberg and autism, see S. Baron-Cohen (2020), "Without such families speaking out, their crises remain hidden," *The Times* (of London), February 9.

29. *Beautiful Young Minds* was later adapted into the dramatic film *X+Y*. You can see an excerpt from the film *Beautiful Young Minds* at "The world of Asperger's," Catalyst, August 28, 2008, www.abc.net.au/catalyst/stories/2346896.htm; on Daniel's diagnosis, see "Beautiful young minds p1," dailymotion, www.daily motion.com/video/x3et56.

APPENDIX 1: TAKE THE SQ AND THE EQ
TO FIND OUT YOUR BRAIN TYPE

1. The SQ-10 and the EQ-10 were developed and tested in Greenberg et al., "Testing the Empathizing-Systemizing (E-S) theory of sex differences."

APPENDIX 2: TAKE THE AQ TO FIND OUT
HOW MANY AUTISTIC TRAITS YOU HAVE

1. The AQ-10 was developed in C. Allison et al. (2012), "Toward brief 'red flags' for autism screening: The Short Autism Spectrum Quotient and the Short Quantitative Checklist in 1,000 cases and 3,000 controls," *Journal of the American Academy of Child and Adolescent Psychiatry* 51, 202–212. It was also tested in Greenberg et al., "Testing the Empathizing-Systemizing (E-S) theory of sex differences."

Figure Notes and Credits

Figure 1.1 Adapted from M. Frank et al. (2017), "Wordbank: An open repository for developmental vocabulary data," *Journal of Child Language* 44(3), 677–694.

Figure 2.1 Courtesy of author.

Figure 2.2 Courtesy of author.

Figure 2.3 Adapted from Rodney Castleden (2002), *The making of Stonehenge* (London: Routledge).

Figure 2.4 Adapted from an image at http://dangerouslyirrelevant.org /2011/11.

Figure 2.5, top Adapted from S. Baron-Cohen and M. V. Lombardo (2017), "Autism and talent: The cognitive and neural basis of systemizing," *Translational Research* 19(4), 345–353.

Figure 2.5, bottom Creative Commons Attribution—ShareAlike 3.0 Unported license. Attribution: Sebastian023.

Figure 2.6 From F. De Waal (2017), "Mammalian empathy: Behavioural manifestations and neural basis," *Nature Reviews Neuroscience* 18.

Figure 2.7 Courtesy of author.

Figure 2.8 Courtesy of author.

Figure 3.1 Courtesy of author.

Figure 3.2 Courtesy of author.

Figure 3.3 From D. Greenberg et al. (2018), "Testing the Empathizing-Systemizing (E-S) theory of sex differences and the Extreme Male Brain (EMB) theory of autism in more than half a million people," *Proceedings of the National Academy of Sciences* 115(48), 12152–12157.

Figure 3.4   Adapted from S. Baron-Cohen et al. (2001), "Studies of theory of mind: Are intuitive physics and intuitive psychology independent?," *Journal of Developmental and Learning Disorders* 5, 47–78.

Figure 3.5   Adapted from S. Ritchie et al. (2018), "Sex differences in the adult human brain: Evidence from 5216 UK Biobank participants," *Cerebral Cortex* 28(8), 2959–2975.

Figure 3.6   Adapted from B. Pakkenberg and H. Gundersen (1997), "Neocortical neuron number in humans: Effect of sex and age," *Journal of Comparative Neurology* 384, 312–320.

Figure 3.7   Adapted from M. Frank et al. (2016), "Wordbank: An open repository for developmental vocabulary data," *Journal of Child Language* 44(3), 677–694.

Figure 3.8   From C. Toran-Allerand (1984), "Gonadal hormones and brain development: Implications for the genesis of sexual differentiation," *Annals of the New York Academy of Sciences* 435, 101–111.

Figure 3.9   Courtesy of author.

Figure 3.10   Created by the author and based on S. Baron-Cohen et al. (2001), "The 'Reading the Mind in the Eyes' test revised version: A study with normal adults, and adults with Asperger syndrome or high-functioning autism," *Journal of Child Psychology and Psychiatry* 42, 241–252.

Figure 4.1   Courtesy of author.

Figure 5.1, top   Photo by Didier Descouens (CC-BY-SA-4.0).

Figure 5.1, middle   Photo by Didier Descouens (CC-BY-SA-4.0).

Figure 5.1, bottom   Photo by Didier Descouens (CC-BY-SA-4.0).

Figure 5.2, left   Heritage Image Partnership Ltd / Alamy Stock Photo.

Figure 5.2, right   From Pierre-Jean Texier, Guillaume Porraz, John Parkington, Jean-Philippe Rigaud, Cedric Poggenpoel, Christopher Miller, Chantal Tribolo, Caroline Cartwright, Aude Coudenneau, Richard Klein, Teresa Steele, and Christine Verna (2010), "A Howiesons Poort tradition of engraving ostrich eggshell containers dated to 60,000 years ago at Diepkloof Rock Shelter, South Africa," *Proceedings of the National Academy of Sciences* 107(14), 6180–6185, https://doi.org/10.1073/pnas.0913047107.

Figure 5.3   Human Origins Program, Smithsonian Institution.

Figure 5.4   Image courtesy Maxime Aubert.

Figure 5.5   Human Origins Program, Smithsonian Institution.

Figure 5.6, top   Historic Images / Alamy Stock Photo.

Figure 5.6, bottom left Landesamt für Denkmalpflege im Stuttgart and Museum Ulm, Photo: Yvonne Mühleis.

Figure 5.6, bottom right DEA / A. DAGLI ORTI/De Agostini via Getty Images.

Figure 5.7 Courtesy of author. This time line is of course schematic, not quantitative in any exact sense. It is meant to convey that we see a small number of simple tools over the past 3.3 million years, but that since 100,000 years ago, the increase in the variety of complex tools has been exponential. Estimates that this growth rate in invention was exponential are based on a time line of human inventions that can be found in a variety of sources: for example, "Timeline of historic inventions," Wikipedia, en.wikipedia.org/wiki/Timeline_of_historic _inventions; "Timeline of scientific discoveries," Wikipedia, en.wikipedia.org /wiki/Timeline_of_scientific_discoveries; "Prehistory to 1650," ScienceTimeline, www.sciencetimeline.net/prehistory.htm; C. Woodford, "Technology Timeline," ExplainThatStuff!, www.explainthatstuff.com/timeline.html; and "Inventions Timeline," www.datesandevents.org/events-timelines/09-inventions -timeline.htm.

Figure 5.8 Photo: Hilde Jensen © University of Tübingen.

Figure 5.9 Courtesy of author.

Figure 5.10 agefotostock / Alamy Stock Photo.

Figure 6.1, Chimpanzee Nature Picture Library / Alamy Stock Photo.

Figure 6.1, Crow Courtesy Dr. Sarah Jelbert.

Figure 6.1, Dolphin Photo: A. Pierini. Dolphin Innovation Project, www.shark baydolphins.org.

Figure 6.1, Octopus Nature Picture Library / Alamy Stock Photo.

Figure 6.1, Firehawk Auscape International Pty Ltd / Alamy Stock Photo.

Figure 6.2 Adapted from D. Povinelli et al. (2000), *Folk physics for apes: The chimpanzee's theory of how the world works* (Oxford: Oxford University Press).

Figure 6.3 Adapted from T. Wynn and F. Coolidge (2012), *How to think like a Neandertal* (Oxford: Oxford University Press).

Figure 8.1 Adapted from H. Witkin and D. Goodenough (1981), "Cognitive styles: Essence and origins. Field dependence and field independence," *Psychological Issues* 51, 1–141.

Appendix Tables 1, 2, and 4 From D. Greenberg et al. (2018), "Testing the Empathizing-Systemizing (E-S) theory of sex differences and the Extreme Male Brain (EMB) theory of autism in more than half a million people," *Proceedings of the National Academy of Sciences* 115(48), 12152–12157.

Appendix Table 3 Created by Varun Warrier and David Greenberg, Cambridge.

Index

© BRIAN HARRIS

Simon Baron-Cohen is professor of psychology and psychiatry and director of the Autism Research Centre at Cambridge University. He is the author of more than six hundred scientific articles and four books, including *The Science of Evil: Zero Degrees of Empathy* and *The Essential Difference.*